The
EVERYTHING
Woodworking Book

I started woodworking about 100 years ago (or so it seems) when I was five years old. I went down into the basement of our house where my dad had a small woodworking shop. I grabbed a handsaw and proceeded to cut some really cool-looking (at least I thought so) slots into the front edge of my dad's bench. Then, to add to my little "project," I found a hammer and some nails.

I pounded and pounded until I had "installed" about a dozen or so nails into the top of the bench.

My dad wasn't as excited as I was about my project, but he didn't discourage me from working in his shop. He got some scraps of wood and showed me how to use his hand tools. I was allowed to go into the shop whenever I wanted.

A few years later, when my woodworking skills had improved, I hand-cut dados, mortises, and tenons in the base of the bench and rebuilt it. I reset the top boards (they were from some old ammunition boxes) and it was like new. For the record, the nails and saw cuts are still in the bench.

I hope that your own adventures in woodworking start out a little more successfully, but end just as happily and productively, as mine have.

Jim Stack

The EVERYTHING Series

Editorial

Publishing Director	Gary M. Krebs
Managing Editor	Kate McBride
Copy Chief	Laura M. Daly
Acquisitions Editor	Courtney Nolan
Development Editor	Larry Shea
Production Editor	Jamie Wielgus

Production

Production Director	Susan Beale
Production Manager	Michelle Roy Kelly
Series Designers	Daria Perreault
	Colleen Cunningham
Cover Design	Paul Beatrice
	Matt LeBlanc
Layout and Graphics	Colleen Cunningham
	Daria Perreault
Composition and Interior Design	Electronic Publishing Services, Inc.
Series Cover Artist	Barry Littmann

Visit the entire Everything® Series at *www.everything.com*

The
EVERYTHING®
Woodworking
Book

*A beginner's guide to
creating great projects
from start to finish*

Jim Stack
and the Editors of *Popular Woodworking*

ADAMS MEDIA
Avon, Massachusetts

Some material in this publication has been adapted and
compiled from the following previously published works:

Churchill, J. *The Woodworker's Complete Shop Reference*
(©2003, Jennifer Churchill, adapted excerpts with permission)

Stack, Jim. *The Biscuit Joiner Project Book*
(©2002, Jim Stack, adapted excerpts with permission)

Stack, Jim. *Building the Perfect Tool Chest*
(©2002, Jim Stack, adapted excerpts with permission)

The World's Best Storage & Shelving Projects
(©2002, Popular Woodworking Books/F+W Publications, Inc.)

Popular Woodworking Magazine, May 2003

Popular Woodworking Magazine, August 2003

Popular Woodworking Magazine, May 2004

ISBN: 1-59337-123-3

Printed in the United States of America

J I H G F E D C B A

Library of Congress Cataloging-in-Publication Data
Stack, Jim
The everything woodworking book / Jim Stack and the editors of Popular Woodworking.
p.   cm. -- (An everything series book)
Includes index.
ISBN 1-59337-123-3
1. Woodwork. I. Popular woodworking. II. Title. III. Series: Everything series.

TT180.S648 2005
684'.08--dc22
                                                                2005009553

Photography by: Al Parrish, Greg DeKraker, Jim Stack,
Bill Hylton, *Popular Woodworking* staff

Illustrations by: Len Churchill, Jim Stack, John W. Hutchinson, Jim Stuard, Melanie Powell

Contents

Part Two: Woodworking Projects, 101

 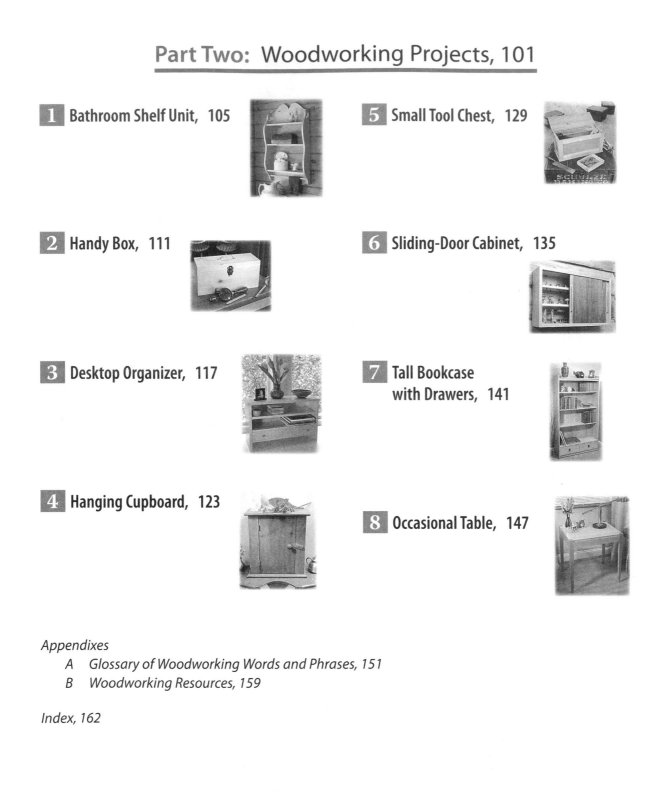

Acknowledgments

The publication of *The Everything® Woodworking Book* would not be possible if it weren't for the following people and companies:

Jim Stack

Popular Woodworking books and magazines

The authors and contributors of *Popular Woodworking* books and magazines—in particular: Jennifer List, Bill Hylton, Mag Ruffman, David Thiel, Jim Stuard, and Christopher Schwarz

F+W Publications, Inc.

Thank you!

Top Ten Reasons You've Decided to Become a Woodworker

1	You'd like to make over your home with something you made yourself.
2	Three simple words: *power tool catalogs.*
3	Over the years, you've watched Norm Abram make twenty-seven cabinets on his TV show *The New Yankee Workshop,* and it's time to build one yourself.
4	You want to follow in the footsteps of such notable woodworking hobbyists as Harrison Ford, Jimmy Carter, and (maybe) your father or grandfather.
5	You just look so darned good in a flannel shirt.
6	You've tried other hobbies, and you found that knitting needles simply don't produce the same impressive roar as a table saw does.
7	Sanding down a tabletop with four different grits of sandpaper is excellent upper-body exercise.
8	You've become addicted to the sweet smell of freshly cut wood.
9	Hanging out at the lumberyard on a Saturday morning is a lot more fun than staying home and cutting the grass.
10	A fine, handmade piece of furniture makes for something more enduring to pass down to your grandchildren than the latest portable music player.

An Important Safety Notice
Please Read

To prevent accidents, keep safety in mind while you work. Use the safety guards installed on power equipment; they are for your protection. When working on power equipment, keep fingers away from saw blades, wear safety goggles to prevent injuries from flying wood chips and sawdust, wear headphones to protect your hearing, and consider installing a dust vacuum to reduce the amount of airborne sawdust in your woodshop. Don't wear loose clothing, such as neckties or shirts with loose sleeves, or jewelry, such as rings, necklaces, bracelets, or earrings, when working with power equipment. Tie back long hair to prevent it from getting caught in your equipment. People who are sensitive to certain chemicals should check the chemical content of any product before using it. You will find further advice on safety throughout this book, and especially in Chapter 5, "Woodworking Safely." Please make sure that "woodworking safely" is what you are doing at all times!

The authors and editors who compiled this book have tried hard to make the contents as accurate as possible. Plans, illustrations, photographs, and text have been carefully checked. All instructions, plans, and projects should be carefully read, studied, and understood before beginning construction. In some photos, power tool guards have been removed to show clearly the operation being demonstrated. Always use all safety guards and attachments that come with your power tools. Due to the variability of local conditions, construction materials, skill levels, etc., neither the authors nor Adams Media nor F+W Publications, Inc., assumes responsibility for any accidents, injuries, damages, or other losses incurred resulting from the material presented in this book.

Introduction

The purpose of *The Everything® Woodworking Book* is to introduce you to the fine craft of woodworking so that you can enjoy creating stylish furniture and cabinetry. You will learn about the many wonderful varieties and qualities of wood, and where and how to buy it. A thorough discussion of both hand and power tools, as well as the various materials you will use in woodworking—hardware, glue, and finishes—will give you all the information you need to put together a well-equipped shop.

We'll then focus on woodworking practices: working safely; choosing the right joinery; setting up your shop; and deciding what to make. In Part Two of the book, with the guidance of helpful step-by-step directions, photos, and drawings, you'll learn how easy it is to build eight practical, attractive woodworking projects.

So dig in! Woodworking is a chance to escape stress, even if just for a little while. While we don't want to exaggerate woodworking's mystical attributes, they are very real—the true spirit and joy that comes from working with your hands, heart, and mind. Wood is one of Mother Earth's greatest building materials, and it is our hope that you will find great satisfaction and peace in your craftsmanship.

—Jim Stack and the Editors of *Popular Woodworking*

Part One

Woodworking Tools, Materials, and Techniques

chapter 1

Wood

Much—maybe even most—of the reason generation after generation of wood-workers has become obsessed with this craft is the remarkable material you get to work with. Even after you've chosen one of the many varieties available—pine? oak? how about bubinga?—you'll find that each individual piece of wood has its own unique color, pattern, characteristics, and feel. As a natural material, wood carries with it the history of decades or even centuries of growth. No other material that you can create beautiful and functional objects from also gives you the same deep connection to the natural world. (This is just one reason, no doubt, you are reading this book, and not, say, something like *The Everything® Plastic-working Book*.)

Because it is a natural material, wood requires special strategies to work with it successfully. You'll find that its habit of expanding and contracting, for example, doesn't stop even after a tree becomes a tabletop. In this chapter, we'll discuss what kinds of wood you can choose from, where and how to buy wood, and what you need to know about wood itself in order to create enduring objects from this irresistible material.

> Types of Wood

More than 100 wood species are available in the United States, with about 60 native woods sold commercially. That's a lot of wood for woodworkers to choose from, and the right choice for a project might not always be the first wood that catches your eye, or the one that is most easily available.

Trees, and the wood they provide, are generally divided into two very broad, and sometimes confusing, classes: *hardwoods* and *softwoods*. The distinction between the two categories is based on the way in which the two types of trees grow. The confusion comes from the fact that there are several softwoods that are actually "harder" than some hardwoods. Southern yellow pine, for example, is a softwood with wood that is harder than basswood, a hardwood.

hardwoods

Hardwoods are deciduous trees with broad leaves. Trees that are hardwoods produce a fruit or a nut and usually go dormant (lose their leaves) in the winter. Hardwoods are also porous, meaning they contain wood cells with open ends called vessel elements that serve as conduits for transporting water or sap in the tree.

Some examples of hardwoods grown in the United States are oak, ash, cherry, maple, and poplar. Most imported tropical woods, sometimes called "exotic woods," are also hardwoods. These imports include some varieties you've probably heard of, such as rosewood, teak, mahogany, and ebony. The names of some other varieties of imported woods may be familiar only to serious woodworkers (and maybe to game-show champions); these include woods such as bubinga, purpleheart, padauk, and cocobolo. For your own woodworking—whether it be furniture making, cabinetmaking, built-in projects, paneling, or architectural woodwork—you will more likely be using domestic hardwoods such as cherry or oak. However, you may want to consider incorporating small amounts of the more exotic (and pricier) imported woods in your projects for such elements as drawer fronts or pulls, decorative strips, door panels, or other accents.

softwoods

Softwoods are generally evergreen conifers, meaning they are cone-bearing. Softwoods are nonporous, meaning they do not contain the vessel elements found in hardwoods.

Softwoods are generally used in the construction of flooring and molding, but also in paneling and cabinetry work. The most common U.S. softwoods available include cedar, fir, hemlock, pine, redwood, and spruce. Most of the lumber you will see at your nearby home center will be varieties of softwoods, especially pine.

Speaking very generally, softwoods tend to be less expensive than hardwoods, as well as lighter. That is to say, if you go to a building-supply store to get a pine board and an oak board of the exact same dimensions, the oak board will probably cost more money and require more effort to lift onto your cart. The comparative costs of all woods—whether hardwoods or softwoods—can vary greatly, though, depending on the particular variety of wood, the quality of the individual piece, and where you are buying it.

If you are a beginning woodworker, there are several reasons you may decide to start off by working with boards of softwoods such as pine. Pine is widely available, is lower in cost than many other woods, and can be easier to cut and shape because of its softness. It is less durable than hardwoods such as oak and cherry, though, and it is more likely to splinter or chip as you are working with it. Hardwoods often have more interesting grain patterns and more varied colors as well.

Softwoods such as pine are a good choice for many projects. They are particularly good when price is a major concern, and for pieces that you are planning to cover with paint, where the wood's grain would be not be visible anyway. You should consider working with hardwoods early on in your woodworking career, though, perhaps to make smaller objects such as boxes. That way you can have the experience of learning to work with higher quality woods without the expense of trying to perfect your skills on large amounts of costly materials.

To learn more about different types of wood, see the photo insert that follows page 86. You'll find a picture and a description of the qualities and common uses for a number of the most common softwoods, domestic hardwoods, and imported woods.

> Moisture, Grain, and Color

The particular ways that trees grow affect the appearance and working qualities of the wood they produce. The part of a tree that a piece of wood comes from and the way the tree was cut up for lumber can also make a big difference.

wood cells and moisture

Wood is composed of long cells that are bound together by a material called *lignin*. When wood splits, it is much more likely to be along the grain—that is, between the wood cells along its length—than perpendicular to the grain in the middle of the board. These differences in the strength of wood along its length versus horizontally will greatly affect your decisions on the types of joints to use for attaching pieces of wood at different angles and for different purposes.

These wood cells also act as sponges, and they can soak up a great deal of water. (In fact, soaking up water is basically a wood cell's job description.) This has a huge impact on what needs to be done to make lumber usable by woodworkers, and it also influences the ways that pieces of wood should be joined together.

▶In order to begin the drying process, this pile of freshly cut planks of cherry is being stickered—that is, laid out with spacers that allow airflow around the boards. The ends of the boards are painted with a special paint that includes wax to keep the ends from splitting.

A live tree contains "free" moisture that is flowing through its cells, and also "bound" moisture within its cells. After a tree is cut down, the moisture within it begins to evaporate, which causes the wood to shrink and often split. To prevent this, logs are cut into planks and "stickered" (that is, stacked with spacers that allow air movement around them). All of the free moisture and most of the bound moisture is allowed to leave the wood, either slowly (often for a year or more) through *air-drying*, or more quickly through a process called *kiln-drying*.

Even after wood is ready to be used, moisture is a concern. With changes in humidity, pieces of wood will shrink or expand. This wood movement occurs almost entirely across the width of a board, and not along its length. When joining pieces of wood together—particularly wide pieces such as a tabletop—this movement has to be taken into account to avoid splitting.

grain patterns

As everyone who has ever seen a tree stump knows, tree trunks usually have a series of rings inside them that you can use to count how many years the tree has been growing. These are caused by the differences in the speed of growth that most trees have between each spring and summer season.

The grain pattern that a piece of lumber has is largely determined by what part of the tree it was cut from, and at what angle it was cut to the tree's annual rings. *Flat-sawn* or *plain-sawn* lumber offers the most common grain pattern, which gives a surface that is oriented at a tangent to the tree's rings. *Quartersawn* lumber is run through the mill so that the surface is perpendicular to the tree's rings. This gives an appealing, consistent grain pattern. *Rift-sawn* lumber gives a grain pattern that is similar on all four sides of the piece of lumber, with a linear grain pattern.

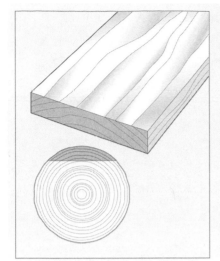
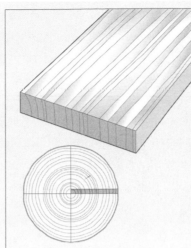
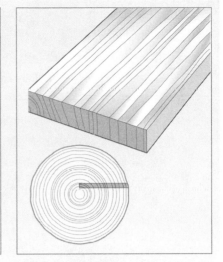

Flat-sawn or plain-sawn lumber   Quartersawn lumber   Rift-sawn lumber

heartwood and sapwood

The part of the trunk that a piece of wood is cut from also influences its color. We always think of cherry as a reddish-brown wood, and of walnut as being a rich, dark brown. The cherry and walnut that come in these darker shades, however, are only from the wood that is found in the center of the tree.

The *heartwood* of a tree is the older and inactive (because the cells are dead) central wood of the tree. The *sapwood* of a tree is the wood surrounding the heartwood. In the living tree, the sapwood carries the sap between the roots and the crown. It is usually lighter in color (often creamy or off-white) than the heartwood. Within a single hardwood board, such as one of cherry, you will sometimes see a section of darker wood from the heartwood next to a section of lighter sapwood. In order to have a consistent color in their work (particularly after pieces are stained and finished), many woodworkers will use only one type of wood, generally the heartwood, within a single project (or even for all their woodworking).

> Lumber Measurements

A quick quiz: What are the dimensions of an 8'-long 2 × 4? How about a 6'-long 1 × 6 pine board?

Yes, *8'* and *6'* really are the lengths of these two pieces. But if you bought those pieces and expected the other dimensions to be *2" by 4"* and *1" by 6"*, you would be in for a bit of a disappointment.

When purchasing softwood lumber, you should know that "nominal sizes" were origi-nally derived from rough lumber dimensions *before* surfacing took place, and so these size listings are always a greater number than the actual dimensions of the lumber. For example, you may think you're buying a 2 × 4, but a strip of wood that may have originally been 2" × 4" is actually surfaced to a final measurement of 1½" × 3½". A 1 × 6 pine board purchased at a home center will normally measure ¾" × 5½".

For all lumber, the standard lengths are in feet: 4, 5, 6, 7, 8, 9, 10, 11, 12, 13, 14, 15, and 16. What you buy at the lumberyard or wood store is referred to as a *board* if the lumber is 1½" or less in its actual thickness, and greater than 1½" in its width. If the piece of lumber is less than 5½" in width, then it may be referred to as a *strip* rather than a board.

board feet

Hardwoods are often sold in less standard sizes than softwoods are. The boards of oak and maple that you find at a home center are still likely to be in exact dimensions such as 1 × 6 or 1 × 10, and in standard lengths such as 6' or 8'. When you go to a true lumberyard, though, hardwood boards will often be sold in variable widths and lengths. The outside edges of boards may be also left rough and unfinished, sometimes even with the bark still attached.

To determine how much hardwood you are getting (and how much you have to pay), a particular system is used. A *board foot* is the equivalent of a board that is 1' long, 1' wide, and 1" thick. This equals, in volume, 144 cubic inches of wood. As an example, the number of cubic inches in a 9' board of cherry that is 1" thick and 8" wide would be the following:

108 [the number of inches in 9 feet] × 1 × 8, or 864 cubic inches

Dividing 864 by 144 gives exactly 6, meaning that
the board would contain 6 board feet

surfaces

Surfaced lumber is lumber that has been surfaced by a machine for smoothness and unifor-mity. If it's referred to as S1S, you're getting a board that is surfaced on one side only. If it's referred to as S2S, you're getting a board that has been surfaced on both sides. If it's referred to as S1E, you're getting a board that has one edge surfaced, and S2E means two edges have been surfaced. *Rough lumber* has not been surfaced, but it has been sawed and edged, often showing saw marks.

For newer woodworkers, one of the most confusing aspects of buying lumber is figuring out the terminology for thicknesses. Rough lumber (which has not been surfaced) is sold in "quarters." Each quarter represents ¼" of thickness in its rough state. So four-quarter lumber (written as 4/4) is 1" thick in its rough state; 5/4 is 1¼", and so on. When the lumber is surfaced by the mill it loses its thickness. That's why 4/4 lumber is ¾" thick when it's surfaced.

> Plywood and Panels

Plywood, which can be made from either hardwoods or softwoods, is basically any flat panel layered with sheets of veneer (called "plies"), joined by pressure and some sort of adhesive. Put simply, it is a structural material made of sheets of wood glued or cemented together. Plywood is always made of an odd number of layers, and each layer consists of one or more sheets of veneer. A veneer is just a thin sheet of wood.

Layers are constructed with the direction of the grain opposite (perpendicular to) one another. The reason for this alternation of grain direction is strength and stability, which means less chance of splitting. Cores may be made of veneer, lumber, or particleboard. Total panel thicknesses are usually at least 1/16" thick and not more than 3" thick. Plywood's main advantages over solid wood are its high strength-to-weight and strength-to-thickness ratios.

types of plywood

Most plywood manufactured for industrial or construction use is produced domestically and is usually made from softwoods such as fir, southern pine, and redwood, although hardwoods can also be used. Hardwood plywood is made from several different species and is usually intended for decorative uses, such as furniture, cabinet panels, and wall panels.

There are two broad categories of plywood: construction/industrial plywood and hardwood/decorative plywood. Most woodworkers are interested in the latter. Softwood veneer may range in thickness from 1/16" to 3/16", and hardwood veneer is often even thinner than that. Then the veneer is transported to clippers that cut it into a variety of widths.

Veneer core plywood is layers of wood veneer sandwiched together with the top and bottom veneers being the best. The lumber core is made of edge-glued boards sandwiched between top and bottom layers of top-quality hardwood veneers. Hardwood edge-banding can be applied after the plywood parts have been cut to size. Another choice is to use iron-on veneer tape. It has heat-activated glue applied to the back of the veneer strips.

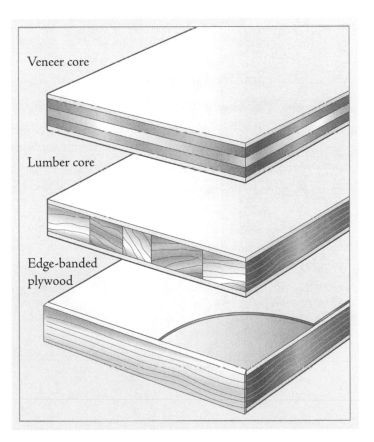

Veneer core

Lumber core

Edge-banded plywood

composition-core boards

Three main types of composition-core veneer boards are available. The cores are made of pulped wood compressed into sheets of particleboard that are sandwiched between a top and bottom layer of top-quality hardwood veneer. The difference between them is the coarseness of the particles used in the cores. Mulch board is the least dense of the three, and medium-density fiberboard (MDF) is the most dense (despite its name). The cost is directly related to the density of the core, with the mulch board being the least expensive. MDF is the choice of many professional woodworkers because of its ultrasmooth surface texture, which creates a perfect surface for fine veneers.

You can also buy various composite boards, such as particleboard, that do not have any veneer facing. These materials are generally used for construction purposes, or possibly for a part of a woodworking project that will not be visible from the exterior.

> Buying Wood

Unless you own a sawmill, finding the best material for your projects is always going to be a challenge. Even professional cabinetmakers are constantly foraging for new sources for wood.

Have no fear, because it is possible to find quality lumber no matter where you live—you just have to know where to look. When you do find some wood, you need to figure out if it's worth buying. Here are some possible sources:

commercial lumberyards

You might not be aware of all the lumberyards that carry hardwoods in your area. Some are small family operations that rely more on word of mouth than marketing. Start by checking your local Yellow Pages and search for "lumber, retail." Visit the WoodFinder Web site *(woodfinder.com)*, which can help you find suppliers within a 200-mile radius.

►Enough lumber for you? Commercial lumberyards like this one are good sources for consistent, graded lumber with few surprises.

Some lumberyards deliver even small loads, and others are worth the drive, so don't discount the stores that are out of town. If you're still having trouble finding hardwoods like red oak and poplar, contact a local cabinet shop and see if someone there can point you in the right direction.

Don't forget to seek out lumber mills if you live near hardwood forests. Some of these mills could sell to the general public—and for reasonable prices, too.

mail-order wood

It might seem crazy to buy wood by mail, especially when you consider that you're buying it unseen and have to pay for shipping, but many of the big mail-order lumber suppliers are quite competitive in price, and the wood is of a high quality.

woodworking clubs

To ease the search for good hardwood, join your local woodworking club or guild. Almost every club seems to have a resident wood scrounger who is more than happy to point you to places that are off the beaten path. Some clubs even organize purchases of lumber for their members.

classified ads, auctions, and off-cuts

There are a few somewhat surprising ways to find wood. Believe it or not, wood shows up pretty regularly in the classified ads of the daily newspaper and the free local shopping papers.

And while you're poring over the classifieds, keep an eye out for auctions at farms and cabinet shops. When these places go under, there can be good deals on wood (and machines). Just remember that haunting auctions is both time-consuming and addictive.

Some people buy lumber through eBay, an online auction Web site. Shipping can be a real killer ($1 a pound), so be careful and do the math before you buy from online auctions.

Finally, for the true bottom-feeder, there's always the waste stream. Find out if there's a pallet factory, furniture manufacturer, veneer mill, or construction site in your area. Their waste, known as *off-cuts*, might be perfect for your woodworking, and especially for your budget.

hardwood lumber grades

When you buy wood at a lumberyard, it has been graded—essentially separated into different bins based on how many defects are in each board. The fewer the defects, the more expensive the board. Grading hardwood lumber is a tricky skill with rules set by the National Hardwood Lumber Association. (Grading softwood is different; these rules do not apply.)

Here are some of the basic guidelines graders follow as they classify each board.

Firsts: Premium boards that are at least 6" wide, 8' long, and 91⅔ percent clear of defects.

Seconds: Premium boards are at least 6" wide, 8' long, and 81⅔ percent clear of defects.

FAS: The two previous grades are typically combined into one grade called *FAS,* or "firsts and seconds," which must be at least 81⅔ percent clear of defects.

FAS 1-face: One face must meet the minimum requirements of FAS; the second face cannot be below No. 1 common.

Selects: While not an "official" grade, this refers to boards that are at least 4" wide, 6' long, and with one face that meets the FAS 1-face requirements. Essentially, these are good clear boards that are too narrow or too short to fit in the previous grades. This and the FAS grades are good choices for nice furniture.

No. 1 common: Boards that are at least 3" wide, 4' long, and 66⅔ percent clear of defects.

No. 2 common: Boards that are at least 3" wide, 4' long, and 50 percent clear of defects.

There are exceptions to these rules. For example: walnut, butternut, and all quarter-sawn woods can be 5" wide instead of 6" wide and still qualify for FAS.

When buying any hardwood, whether it has been graded or not, it is best if you can pick through the selection of boards yourself to find the ones that are the highest quality and the best for your projects. In some situations—such as when you are ordering wood to be delivered through the mail—this may be impossible, but in all other circumstances, always ask if you can select your own boards.

> Wood Talk

Whichever kind of wood you decide to use for your woodworking projects, there are a number of terms and common practices you will need to understand before heading down to your local home center or lumberyard. Some of the following terms have already been discussed, some not, and you'll find some more terms related to wood in the glossary on page 151.

air-dried lumber: Wood that has been dried from its freshly cut state by stacking it (usually outside) with stickers between. (Air-drying reduces the moisture content to about 12 to 15 percent. Wood for interior use needs to be dried further.)

board foot: A piece of wood that is 1" thick, 12" wide, and 12" long in the rough, or its cubic equivalent.

chatter mark: A defect caused when the board was surfaced at the mill and the knives mar the surface.

cupped: A board with edges higher than its middle. (The cup is always to the sap side of the board.)

defect: An imperfection in the board that will change how it is graded (and its price).

dimensional lumber: Lumber that is surfaced on all four sides (S4S) to specific thicknesses and widths: 1 × 4s, 2 × 8s, etc. (Note that with this lumber the finished thickness and width are less than the stated size. For example, a 1 × 4 typically will measure ¾" × 3½".)

end check: A split or crack caused by the separation of the wood fibers at the end of a board, almost always as a result of drying.

flitch: When a log is sawn into veneer and the sheets are stacked in the same order in which they came off the log.

green lumber: Wood that has been freshly cut from the tree, typically with a moisture content of 60 percent or higher.

heartwood: The part of the tree between the pith (the very center) and the sapwood (the whitish outer layer of wood).

honeycomb: A separation of the wood fibers inside the board during drying. (It might not be evident from the face of the board.)

kiln-drying: An artificial way to reduce the moisture content of wood using heat and forced air.

knot: A circular woody mass in a board that occurs where a branch or twig attached to the tree.

lineal feet: A measurement of wood that's 12", regardless of the board's width or thickness. (usually refers to moldings)

mineral streak: A typically green or brown discoloration, which can be caused by an injury to the tree.

moisture content: The percentage of a board's weight that is water.

pitch: A resinous, gummy substance typically found between the growth rings of softwoods.

pith: The small and soft core of a tree that the wood grows around. (It's undesirable for woodworking.)

plain (flat) sawn: A method of milling a log that results in the growth rings intersecting the face of the board at an angle less than 45°.

quartersawn: A method of cutting a log at the mill that results in the growth rings intersecting the face of the board at more than 45°. (Quartersawing wastes more wood and requires more effort. But quartersawn wood is more stable.)

random widths and lengths: Hardwoods that are cut into different widths and lengths to get the best grade. (While softwoods and cabinet woods such as red oak and poplar can be found as dimensional lumber, many hardwoods cannot.)

rift sawn: A method of cutting a log that results in the growth rings intersecting the face of the board at an angle between 30° and 60°. (It is more stable than plain-sawn wood and less stable than quartersawn.)

rough: A board as it comes from the sawmill; not surfaced or planed.

SLR1E: The acronym for "straight-line ripped one edge," meaning the board has one true edge.

S2S: Planed on two faces; the edges are rough.

S3S: Planed on two faces and one edge; one edge is rough.

S4S: Planed to a smooth finish on all four long edges of a board.

sapwood: The lighter-colored wood between the heartwood and bark. (It is typically weaker than the heartwood.)

shake: A split that occurs before the tree is cut, typically from the wind buffeting the tree.

shorts: High-quality lumber that is less than 6' long.

sound knot: A knot that is solid across the face of the board and shows no sign of decay.

straight-line rip: A perfectly straight edge that is suitable for gluing.

surface check: A shallow separation of the wood fibers.

twist: Where the board has warped into a spiral.

wane: The presence of bark on the edge or corner of a piece of wood.

warp: A general term for a distortion in a board where it twists or curves out of shape.

worm holes: A void in the wood caused by burrowing insects (killed during kiln-drying).

chapter 2

Tools

We mentioned at the beginning of Chapter 1 that perhaps the greatest attraction in woodworking is the wood itself. A close second for many woodworkers is the enticement of the many wonderful tools that you get to use. Maybe you'll go the traditional route and enjoy the sweet sound of a hand plane as it shaves off an even-thinner-than-paper strip of wood. Or you may go a more modern (and noisier) way and marvel at how easily you can use a router and a jig to quickly cut all the dovetail joints you need for a chest of drawers. Most likely, you will find that for many jobs power tools can save you time and effort, while for other tasks hand tools are often the best choice. There is no one best way to practice woodworking, but there is a nearly endless array of tools to choose from to create your own path for turning a pile of lumber into a prized creation.

> Hand Tools

When you're out shopping for your shop, it's easy to get mesmerized by all of those modern power tools. You may get the idea that hand tools are simply relics of a long-gone era, and that power tools are the only way to go for quick, efficient woodworking. But just look at the intricate, high-quality furnishings that were produced in the seventeenth, eighteenth, and nineteenth centuries. Those craftsmen worked without a power tool to their name. Some of the earliest woodworking techniques are still some of the best.

A word to the wise: Just because hand tools aren't plugged into electric sockets doesn't mean they're safe. Be cautious—watch and learn from experienced woodworkers how to confidently and properly handle all of your woodworking tools. Never lose respect for the tools you're using and never forget that hand tools can be just as dangerous as power tools. We'll discuss proper safety for both types of tools further in Chapter 5.

hand planes

For some reason, hand planes are one of the most sensually pleasing woodworking tools to hold and to use. The shape of the tool, contoured to your hand, and your firm grip on it make you feel like you're in charge of your woodworking project while keeping you closely connected to it. You can feel and hear the wood beneath your hands as it peels off in (what you hope are) even, level strips. Handsaws are similar in this manner.

So what are hand planes used for? And what are your hand plane options?

In general, hand planes are used to smooth the surface of a piece of wood. The plane's body, made of metal or wood, holds a metal blade at a particular, fixed angle to the wood and is drawn across the wood in a scraping motion to "true up" the surface. Hand planes have adjustments on them to change the depth of the blade's cutting edge.

Shaping planes are a little more intricate, as they are used to shape contours. Hollow planes, in particular, can be used to hollow out things such as bowls.

▶ Nothing evokes the long tradition of fine woodworking better than a collection of hand planes.

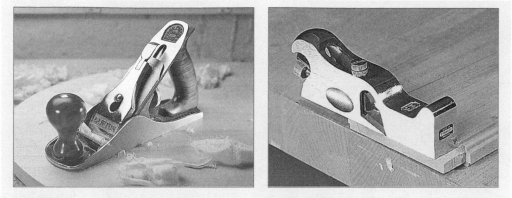

Bench plane   Shoulder plane

You can buy some pretty fancy hand planes for up to hundreds of dollars, but you also can't go wrong with an old Stanley hand plane that you inherited from your father or find at an antique store (after a little restoration work to remove the rust, of course). As with fine musical instruments, many aficionados feel that hand planes are one object where the technological developments of the last few decades have not necessarily led to a better result.

handsaws

First, know that handsaws are great exercise. The power comes from your own hand (hence, the name), unlike the electric power of a circular saw or a miter saw, for example. When it comes to saws, you'll find that the fewer teeth per inch (TPI) on the blade, the faster and rougher the cut will be. More TPI means you'll get a slower but more precise cut.

Crosscut handsaw

Some handsaws are made for very particular tasks, such as cutting tight curves (coping saws) or cutting intricate dovetail joints (the gent's saw). But you'll most often come into contact with these basic types: the crosscut saw, the ripsaw, and the Japanese ryoba saw.

Ripping handsaw

Crosscut saws are used for cuts that you want to make across the wood grain. Crosscut saws usually come with 8 to 12 TPI.

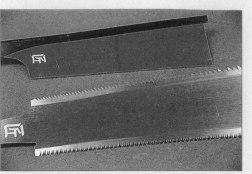
Dozuki saw (top) and ryoba saw (bottom)

Ripsaws are used for cuts that you want to make with the wood grain. Your cut will be rougher than with a crosscut saw, as ripsaws have fewer TPI, usually from 5 to 7.

Japanese saws work on the pull and have narrow kerfs (that is, they take away only a narrow slice of wood with each cut). A kataba-nokogiri (*nokogiri* means "saw") is a saw with teeth on one side; a ryoba-nokogiri is a saw with teeth on both sides (one edge for ripping and the other for crosscutting); and a dozuki-nokogiri is a saw with a stiffener.

Ryoba saws are used for general cutting purposes. The kugihiki saw is great for cutting things flush to the surface, such as cutting a dowel peg overhang flush to its surface. The dozuki is good for cutting dovetails and other types of joinery.

scraping and smoothing hand tools

Scraping and smoothing hand tools are used by general woodworkers, but also are often utilized by specialty woodworkers, such as woodturners and woodcarvers. If you work in those two latter fields, you may have a much bigger arsenal of gouges and chisels than the average woodworker.

But if you're a novice, you want at least the following basics.

chisels

With big, thick blades, chisels are a necessity in any woodshop. Most obviously, they can be used to chunk out big pieces of wood from a cut. Chisels are also used to quickly shape and smooth wood when cutting mortises or other joints. At your local hardware or home-improvement store you'll see a wide variety of chisels in many sizes, usually ranging from ⅛" to 2", and with mostly self-explanatory names, such as paring (or butt) chisels, framing chisels,

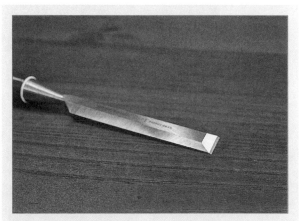
Finely sharpened chisel

mortising chisels, and crank-neck chisels. Gouges, which fall into the chisel family but have been modified to include a hollowed blade, are used for roughing out stock or gouging out

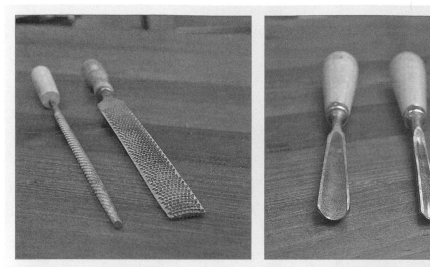

Round or rat-tail rasp (left) and combination flat and curved rasp (right)

Gouges come in many sizes.

hollows into wood surfaces. A wooden mallet is sometimes used to drive a chisel into a cut without damaging the chisel's handle.

rasps and files

Rasps are used for shaping wood wherever a chisel might be too large and clumsy or might create tear-out in your project. You'll find rasps in flat, half-round and round shapes and in a range of coarsenesses. Files are basically rasps but offer a finer cut.

A homemade wooden mallet

clamps

As every experienced woodworker knows, you can never have enough clamps.

►These clamps are, left to right: a deep-throat quick clamp, a spring clamp, a lever-operated quick clamp, a one-hand-operated quick clamp, another style of a one-hand-operated quick clamp, a combination quick clamp with screw tightener, two styles of quick clamps with screw tighteners, and a C-clamp.

Clamps do as their name implies: They clamp elements of an assembly together to hold the shape, to hold it secure to a worktable, or to hold pieces together in place during glue-up. Some of the most common clamps are:

C-clamps: These small, powerful clamps are useful for smaller projects.

Pipe clamps: These inexpensive clamps are versatile and simple.

Spring clamps: These quick, easy-to-use clamps are like giant clothespins; they're simple, but they don't work for heavy situations.

Hand screws: These clamps, which have been around for ages, are made of wood that won't mar your projects, have deep throats, and can be adjusted to different angles.

Miter clamps: These clamps hold mitered frames or joints together.

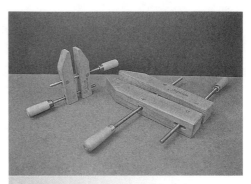

▶Hand screws are great for gluing angled joints because the jaws can be adjusted out of parallel.

measuring and marking tools

From the rudimentary chalk line to the bevel gauge, quite a few hand tools are available to assist in the measuring and marking of your projects. Of course, every woodworker has a good tape measure, probably even two or three of them. Even most nonwoodworkers own at least one tape measure. It's probably the most utilized device in or out of the workshop.

Tape measures are good for measuring long boards without hauling a straightedge around or a folding tape. Most tape measures advance in increments of either $\frac{1}{16}$" or $\frac{1}{32}$".

The standard 1' ruler, yardstick, and 3' straightedge are other items you will likely get a lot of use out of in your shop. All of these can be used for measuring and/or drawing straight lines.

The most versatile and essential measuring devices are the 25' or 30' measuring tapes, the 6" and 12" speed squares, and the movable-angle gauge or sliding T-bevel, which is an incredibly versatile tool that allows you to precisely find the exact angle you need to make an odd angle.

Another unique tool is the center square, which can be used to find the center on a round object. Simply place it on the end of a dowel, for example, line it up to the two pins underneath, make a crosshatch each way, and there you go: there's the center.

Outside calipers are used most of the time for measuring thicknesses of items that are being turned on the lathe. They also seem to work excellently for spindle replacement on new furniture; you can use them to measure the peg that will go into the seat or the backrest.

◆ **Tape measure:** used for quick measures and measuring accurate lengths

◆ **3' straightedge:** used for drawing straight lines and accurate measuring

◆ **90°-angle device:** used for measuring accurate angles

◆ **Angle gauge:** used for measuring angles and gauging angles

◆ **Compass:** used to divide lines and measure circles and arcs; also used to create polygons

◆ **Protractor:** used for measuring angles and lines

◆ **Carpenter's square:** used as an adjustable slide for accurate measures across, at 45° angles and at 90° angles

◆ **Speed square (6" and 12"):** used for accurate straight lines, as well as rafter cuts and layouts

◆ **T-square:** used for larger square layouts

◆ **French curve:** used for laying out intricate designs for "fancier" projects

◆ **Outside calipers:** used for measuring the thicknesses of items being turned on the lathe

> Power Tools

Woodworkers and their power tools. It's almost not necessary to outline what types of power tools are on the market, as most novice woodworkers have an intimate relationship with the tool aisle at the local home-improvement center before they even attempt their first project. For many woodworkers, it's all about the tools.

While the old-fashioned and stubborn may insist that you can build anything with hand tools that you can build with power tools, it's definitely true that power tools save a lot of time and a lot of sore muscles.

But you need to acquaint yourself with what's out there and what your woodworking needs will be before you go out and spend a fortune on very nice, but possibly very frivolous, power tools.

types of power tools

It's unarguable that power tools make a woodworker's life easier. They also make it more expensive—unless you consider time to be money, because power tools definitely save the former.

The kind of woodworking you'll be doing dictates which tools you absolutely must have in your shop. A great overview of all the tools you might need is *The Insider's Guide to Buying Tools* (Popular Woodworking, 2000); it discusses in detail the uses and pros and cons of everything from air tools to biscuit joiners.

If you're going to be making furniture, you almost have to have a table saw. Which, of course, means you almost have to have a dust collector. From there, it becomes more individualized. Power tools are available in two general categories: portable and stationary (also called "floor models"). There are miter saws, drill presses, jigsaws, jointers, planers, routers, scroll saws, and more. And, if you plan to do your own spindle turning or other turning, you'll need a lathe.

circular saws

If you currently own only one power saw, there's a good chance it's a circular saw. This makes some sense, as a circular saw is one of the most flexible power hand tools available. This saw can be used to crosscut, rip, cut angles by tilting the base, cut dadoes, make plunge cuts (for making cutouts for sinks in countertops, for example), and more. An entire house could be constructed using this tool as your only saw. Different blades are available depending on what materials you need to cut.

Circular saw

The more expensive saws have a worm drive connecting the motor to the blade rather than the traditional direct drive. The worm drive has proven itself to be effective in promoting motor life. It is recommended that a heavy-duty extension cord be used with these saws.

table saws

While circular saws are tremendously versatile and portable, and relatively inexpensive, they really are more suited to carpentry than to fine woodworking. For most woodworkers, the table saw is the single most important tool in the shop. If you have one, it is definitely the centerpiece around which the rest of your shop revolves. Unless you're going to be crosscutting and ripping all of your wood with a circular saw, you won't go far without a table saw.

Table saw

Table saws are used not only for ripping and crosscutting, but also for making joints by cutting dadoes, tenons, and other elements, all at a variety of angles.

Table saws are only as good as the blades in them; sharp, clean blades are essential. And you'll need different blades for different tasks. For example, ripping blades, which generally have large teeth and the fewest teeth per inch (TPI), are used (obviously) for ripping. Crosscutting blades, which crosscut solid-wood stock, have smaller teeth and about 60 TPI. Dado blades are used for cutting grooves wider than a regular single blade can cut.

band saws

Band saws are available in both portable and stationary models. They're most commonly used for following template curves or for cutting wood into usable sizes. What makes the band saw unique is its continuous, flexible steel blade, which comes in a variety of widths and tooth grinds.

miter saws

Typically, miter saws are used for cutting angles. Your choices include conventional miter saws, compound miter saws (which

Band saw

allow for more complex cuts), and sliding-compound miter saws (which offer the most

Miter saw

intricate crosscut capabilities). The conventional miter saws basically cut 45° angles, while the more complex and more expensive compound miter saws cut a variety of angles. Pay attention to the types of blades available; they can be specific to the type of cut you desire in a particular task or project.

jigsaws

Jigsaws are handy little tools that come in both corded and cordless forms. They're kind of the portable version of the band saw. What they lack in power, jigsaws make up for in versatility. They can cut through wood, metal, and plastic, depending on the type of blade you have inserted. Jigsaws are great for cutting circles or curves.

Jigsaw

drills

Drills are essential tools that have infinite uses even outside of the woodshop. Like other portable tools, they come in corded or cordless varieties. Drills are mostly used for drilling holes or attaching fasteners. Like routers, drills come with a wide range of accessories, such as jigs, drill guides, and drill bits. Although drills come in different sizes, the ⅜" drill is the most common in woodshops.

routers

So many options and uses are available for routers that entire books have been written about this versatile tool (see page 76 for a photo of a router in action). There are standard (or "fixed-base") routers and plunge routers. They can be used to create moldings, mortises, and wide grooves. You can choose from a vast array of router accessories, depending on your needs. These include an infinite variety of router bit attachments, foot switches, jigs, speed controls, template guides, and more.

biscuit joiners

To some, the biscuit joiner is a magical tool that allows quick assemblies of projects and strong joints. The joiner tool cuts slots into adjoining pieces of wood, into which a wooden biscuit (after being dipped in glue) is inserted, forming the joint attachment. More and more, once skeptical woodworkers are being won over by the ease and simplicity of the biscuit joiner.

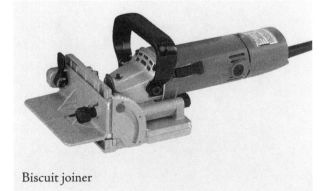

Biscuit joiner

drill presses

Portable (or benchtop) as well as stationary drill presses are available. Used for drilling holes at specific angles and depths, drill presses come in a wide variety of sizes. Different drill bit attachments are available, such as Forstner bits (which cut very neat, flat-bottomed holes and also can be used to enlarge existing holes) and plug-cutter bits (which cut tenons or up to 1"-diameter plugs for covering counterbored screws).

jointers and planers

Jointers and planers come in both benchtop and stationary models. The individual stationary models are large, expensive machines that take up quite a bit of shop space; if space is an issue for you, combination jointer/planer models are available. Basically, planers have metal cutters to smooth out rough-sawn lumber, making both sides of a board flat and parallel to one another. Jointers offer a continuation of this board preparation process and take the place of the sweaty exercise involved in hand-planing the faces and edges of boards to create a smooth surface. They are extremely useful if you are joining (hence the name) the long edges of two boards together, as in a tabletop, and you need the edges to be perfectly flat and straight in order to avoid even the tiniest of gaps.

lathes

Lathes also come in portable and stationary models. If you'd rather not take the easy way out and buy preturned legs and spindles, then you'll need at least a small benchtop lathe. Bigger stationary lathes are more in the realm of serious furniture makers and of woodworkers who focus mostly on decorative turning, creating delicate wood pieces such as vases and bowls. Basically, a lathe supports a plank of wood and turns it, then the woodworker uses a tool such as a gouge against the quickly spinning wood to create the desired size and shape.

Lathe

sanders

The wide variety of sanders includes random-orbit, belt, oscillating-spindle, stationary, and more. Each sander has a different application. Stationary sanders, such as belt/disk sanders, are so powerful that they not only sand wood smooth, they can flatten and square it. Portable hand-held sanders are more common in the individual woodshop. They're used to shape and smooth project surfaces and edges.

Spindle sander

dust collection systems

Dust collection is about your own health and about shop cleanliness. You'll never escape the fact that woodworking produces wood dust. But with a good dust collection system in place, you can reduce your risks of lung or breathing problems. Each tool needs its own system attached, but you can also invest in a simple wet-dry vacuum or an air cleaner.

> Sharpening Your Tools

In a survey asking woodworkers for the Top Ten Fun Things to Do in Woodworking, sharpening tools came in thirteenth. Okay, we made that up, but this much is true: Using sharp tools makes woodworking go faster and brings better results, but getting tools sharp is not many people's idea of a good time. In the following sections, we'll introduce some of the procedures and materials used in tool sharpening. For more information on how to get your tools razor-sharp (actually, probably sharper than any razor you have), consult a book with more detailed discussion such as *The Complete Guide to Sharpening* (Taunton, 1996).

power grinding

Sharpening tools is a fairly precise science, involving proper angles and bevels. It can also be a lot of work, so it's not a surprise that the world of power tools has made its way into the sharpening arena.

Grinding wheels, also called grinders, are fun if you like to see sparks flying and fantasize that you're really a welder for a few minutes. They also can save you a lot of time and energy if you use them carefully.

Basically, you want to pay attention to grit and grain when buying a grinding wheel. Grinding wheels have thousands of little, abrasive grains distributed throughout the surface, which move against the tool to cut away metal. This is the sharpening process. Grinders are best for removing a lot of metal at once, when you get big nicks in your blades, for example. To make your tool sharp, you'll still have to hone the blade after shaping the tool's edge on the grinder.

Dry grinding wheels are quite common in today's woodshop and are used for sharpening edge tools. Wheels on grinders are usually made of Carborundum (silicon carbide) or aluminum oxide. In general, you'll find Carborundum wheels on hand-held grinders. Aluminum oxide wheels are used on benchtop grinders. Grinding wheels are available in a variety of shapes, sizes, and grits.

Don't burn your tools! One thing you need to be careful of when using a grinding wheel is the tendency of a dry grinding wheel to overheat and burn the metal of your blade.

▶ This wet-wheel power grinder comes with two different wheels (one for grinding and one for honing), variable speeds, and a guide you can set to grind a blade at the proper angle.

Pouring water repeatedly on your tool to cool it down can damage it (by causing it to crack), so it's recommended that you place the hot edge on a piece of metal to cool slowly.

You can buy wet-wheel grinders, as well, which are slower and more expensive, but eradicate the worry of overheating your blades.

Some manufacturers recommend that you don't use dry grinding wheels to sharpen your quality chisels or plane irons made of carbon steel. Wet wheels grind more slowly, but are kinder to your tools.

Woodworkers differ in their opinions about whether it's best to grind on a grinding wheel (known as grinding the bevel on a blade hollow) or to manually grind flat on a grindstone.

sharpening stones

While grinding wheels are nice and will save you some labor and time, stones and files are also capable of sharpening your blade edges to incredible sharpness.

The options are almost infinite. Sharpening stones come in a wide variety of sizes, shapes, and coarsenesses. First, there are waterstones and oilstones. Sharpening-stone oil is a special mineral-based oil used to lubricate the sharpening stone and flush residue from its surface. This residue, which is comprised of a variety of materials, including the metal from the tools being sharpened, is called *swarf.*

▶ The first step in sharpening a new chisel is to flatten its back, as is being done here with a sharpening stone. The wooden box that holds the stone makes it easier to clamp the stone to a workbench so that it stays firmly in place during the sharpening process.

Woodworkers are probably best served by using waterstones, as oilstones are messy and more time-consuming to use. But be very careful to thoroughly dry your tools after using waterstones so they don't rust as a result of the moisture.

Waterstones are made of either aluminum oxide or silicon carbide and range in coarseness from 250 grit to 8,000 grit.

In your sharpening arsenal, you'll want to have a 600- to 800-grit water-stone, a 1,000- to 1,200-grit waterstone, and a 6,000- to 8,000-grit waterstone in order to have a full complement of coarsenesses for different tools and sharpening tasks.

Although rather expensive, diamond stones (which have diamond dust embedded into the metal) last a long time and produce quality edges. They come in coarse, medium, fine, and extra-fine coarsenesses. You can use fine-grit diamond stones for sharpening carbide-tipped tools.

Ceramic stones are another option. Less expensive than diamond stones, ceramic stones are made of aluminum oxide and are great for sharpening high-speed steel.

> Buying Tools

Not too long ago buying a woodworking power tool meant driving to Sears, the local hardware store, or—if you were lucky—a woodworking machinery store. You could look at the Craftsman tools, or check out the two or three drills the store stocked and a couple of different drill presses.

There wasn't a huge decision to be made over price because your choices were limited. You could do some shopping by mail, but the depth of selection wasn't very impressive.

Today, you can compare prices on more than 100 different cordless drill models, find the best deal, and buy the tool—all without moving from your desk. And that goes for big items like cabinet saws and 20" planers, as well.

The tool-buying revolution that has been created by competitive catalog and mail-order sales, and more recently the advent of Internet shopping, puts an enormous amount of tools at our fingertips. In most cases this is a good thing, but there are a few pieces of information that you should be armed with before picking up the phone so that you save yourself time and hassles.

know what you need

With so many tools and options available, having good information is especially critical. First, figure out what type of tool you need and what features are most important to you. There are likely to be a couple of tools from different manufacturers that will fit your needs, so take a look at the price ranges, and choose one or two that you prefer.

Another way to get specific tool advice is to reach out over the Internet to other tool shoppers and owners. At Internet discussion groups such as WoodCentral *(www.woodcentral. com)* and Usenet.com newsgroups *(rec.woodworking* at *www.usenet.com)* you can ask other woodworkers about specific brands and models. In many cases, these discussion groups archive their messages. That's a good place to start looking for tool advice. Simply asking everyone in the group, "Which table saw should I buy?" is likely to invoke a lot of varied responses. Research what has already been said about the tools you're interested in, then ask your question.

Other Internet sites offer new ways to get tool-shopping information as well. Productopia *(www.productopia.com)* offers product reviews and makes recommendations on purchases for a number of tool categories. They also include links to the manufacturers' sites as well as shopping sites.

Another twist can be found at *www.mercata.com.* This site offers tools (as well as a lot of other products) for sale in a group-buying deal for lower prices. The tool you're looking for isn't always for sale, but there are some bargains to be had.

where to shop for the best price

Next, dig out your catalogs or log on to the Internet to check the prices on the tools you chose. You should be able to find a tool that fits in your price range—but don't buy it yet. First, consider where you can shop and what each location has to offer.

Most woodworkers have a love-hate relationship with home centers. They love the convenience, selection, and price—but it can be difficult to find someone who can help you choose a tool intelligently.

Many smaller tool stores used by professionals have a knowledgeable staff that can help you pick the right tool. And you can still take your purchase home immediately. Unfortunately, this personal touch and immediacy comes with a little higher price. If you choose to spend time with a knowledgeable tool salesperson, don't abuse his or her livelihood by then heading to the Internet for the cheapest price. If you use the information, pay for the privilege.

Catalogs and the Internet offer a simple way to find the best price on a tool if you've already decided what to buy. If you need advice, you can be out of luck. Some newer sites and catalogs offer buying information, but at many sites it's superficial and spotty. Also, check to see how long it will take you to get your tool. If it's a cordless drill you stand a good chance of getting it within a week. Larger machinery can take around ten days. And be sure to check the shipping charges and make sure you include them in your price comparison. Smaller tools may ship for less than $5, but depending on where you buy, shipping a 15" planer can cost you up to $100.

When remote shopping, make sure you check the return policy of the company. While most reputable merchandisers will take back a damaged or nonworking tool without question, they may only be able to issue a store credit rather than return your money. They may also have a time limit on how long you have to send the tool back. There is also the possibility of a restocking charge of 15 to 20 percent. Don't expect trouble with a return, but make sure you are aware under what terms you're buying your tool.

Also be sure to factor in the hassle you may face with packing a mail-order return. Sending a jointer back across the country because its beds are warped is a lot harder than taking it across town to a local distributor.

When buying from any merchandiser that is shipping your product, check the delivery instructions. They may use a carrier that requires a signature to release the package. If you're not home—no package. In the case of larger items you need to be aware that your table saw will likely appear at your door on the back of a tractor trailer. How you get it off the truck and into your house is your problem, not the driver's. Anticipate the problem.

Fasteners, Hardware, and Glues

All right, now you've got the wood, and you've got some tools to cut, shape, and smooth it. How are you going to get those pieces of wood connected into a fine piece of furniture?

It has been said that, at the most basic level of working wood, metal fasteners are what holds the wood together. At one step up, a bond of glue is what connects the wood; and at the highest level, it is the wood itself that keeps a piece from coming apart. In Chapter 6, when we discuss joinery, we'll see the various ways that you can shape wood into enduring joints that approach this higher level of craft. At all stages of your woodworking, though, you'll find that fasteners, hardware, and glues are important elements to consider. Different types of fasteners and varieties of glue work best for different functions and in different circumstances. In the case of hardware, your choice of a particular hinge or drawer pull can affect both how you make a piece and what style it will have.

> Metal Fasteners

In carpentry, metal fasteners are an essential and unapologetic method for attaching one piece of wood to another. In woodworking, such fasteners as nails, screws, and (rarely) bolts have their place, though it is usually a more hidden one—metal screw heads that are set below the surface and covered by wooden plugs, or brads that are nailed on the underside of a drawer to keep its bottom from sliding out. As with everything else in woodworking, the goal is always to find the right tool (for that is what fasteners really are) for the right job.

nails

Nails are used mainly to join or fix pieces of wood together, usually softwoods, as hardwoods often cause the nails to bend under the hammer's impact. If you do need to nail into a wood as hard as oak, for example, you will often need to drill a small pilot hole first, but generally speaking, nails aren't the best fasteners for hardwoods.

You should never wax or oil nails to get an easier drive into the wood. Friction is needed between the nail shaft and the wood in order for the nail to hold, and wax or oil will minimize the friction. Instead, drill a pilot hole half the diameter of the nail's shaft, then pound in the nail.

Woodworkers concern themselves with basically four types of nails: *common, casing, finishing,* and *brads* (which are tiny finish nails).

Most common are, of course, the common nail and the finishing nail. Common nails have large, flat heads and are used for "unpretty" work that can look fairly rough. If a woodworker wants a prettier final project, the thinner finishing nails are a better choice (mainly because you can drive the small head deep into the wood and fill the gap with wood putty).

When shopping, note that nails are sold by head type and by pennyweight. The penny system can be a little confusing. Originally, it referred to the weight of nails per hundred.

Here is a helpful place to start in understanding it: A 2-penny nail (signified as "2d") is a nail that is 1" long. Any nail over 20d is called a *spike*. Or you can try this little formula: For nail sizes that run up to 10-penny (signified as "10d"), you can figure out the nail's actual length by (1) dividing the penny size by 4 and (2) adding ½" to that number. For example, a 4d nail would be 1½" long.

Penny Size vs. Nail Length		
2d	–	1"
3d	–	1¼"
4d	–	1½"
5d	–	1¾"
6d	–	2"
7d	–	2¼"
8d	–	2½"
9d	–	2¾"
10d	–	3"
12d	–	3¼"
16d	–	3½"
20d	–	4"

And it works the other way, too. If you know the length of the nail you need and are trying to figure out which penny nail to buy at the hardware store, use the previous formula in reverse: (1) subtract ½" from the length of the nail and (2) multiply that number by 4. So, if you know you need a 2½" nail, subtract that ½" and multiply 2 by 4. Since 2 times 4 equals 8, you now know you need an 8d (8-penny) nail. Another way to sort this all out, if you're better at memorization than at mathematics, is to learn the penny sizes and lengths in the Penny Size vs. Nail Length chart.

screws

Although there are four general categories of screws—drywall screws, production screws, wood screws, and sheet-metal screws—woodworkers are usually interested only in the latter two types, wood and sheet-metal.

The main difference between wood and sheet-metal screws is the thread (the part on the shaft of the screw that digs itself into the wood). Whereas sheet-metal screws can, obviously, penetrate metal surfaces better, due to sharper threads, wood screws have a shallower thread designed for softer materials. Both types of screws are sold in the same general measurements.

Wood screws often have bright zinc plating, but are also sold as galvanized, solid bronze, brass, stainless steel, or aluminum. They're sold almost everywhere.

Sheet-metal screws can be used to attach hardware in the woodshop, or to attach anything to metal.

You also may encounter medium-density fiberboard (MDF) screws, which do not have tapered bodies like wood screws, but have sharp threads useful in particleboard and MDF projects.

Screws usually are sold in lengths ranging from ¼" to 3" or 4". The diameter of the screw is referred to by a gauge number. The combinations of screw lengths and gauges that you can buy are almost infinite. You would order a screw like this: "Please give me a ¼" No. 3 screw." Here the ¼" refers to the length, and the "No. 3" refers to the diameter size.

Screws numbered 0 are 0.060" in diameter; No. 1s are 0.073"; 2s are 0.086"; and so on, up to screw number 24, which is 0.372" in diameter.

Screw heads also come in a variety of shapes: flat, round, and oval. Flat-head screws are the ones used most commonly in woodworking because of the flush surfaces they provide.

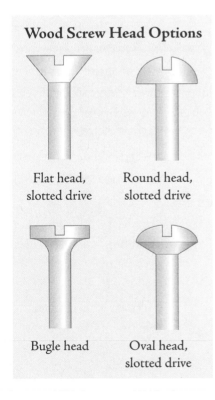

Wood Screw Head Options

Flat head, slotted drive

Round head, slotted drive

Bugle head

Oval head, slotted drive

Screw heads also come in a variety of *drives,* which each require a particular type of screwdriver. Everyone is familiar with the conventional slotted drive and the Phillips; a few other options are shown in the illustrations.

If you need to drive in a large number of screws, especially longer ones, you will find it worthwhile to buy a bit for your electric drill in the drive appropriate for the screws you are using (most often, this would be a Phillips-head bit). If you don't want your screw heads to be visible in your finished project, you will also need to countersink them. To do this, you use a drill bit to make a shallow, v-shaped hole slightly larger than the screw head. After you have driven the screw into the hole created, you can cover it with a wooden plug that you can cut or sand flush to the wood's surface. Combination bits are available that can drill a pilot hole into the piece at the same time that they make the hole for countersinking the screw.

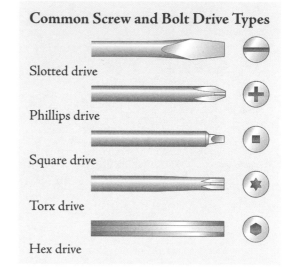

Common Screw and Bolt Drive Types

Slotted drive

Phillips drive

Square drive

Torx drive

Hex drive

bolts

A *bolt* is basically a metal rod that fastens objects together. It has a head at one end and a screw thread at the other, and it's secured by a nut.

Bolts aren't often used in fine furniture projects, or in many woodworking projects in general, but they do show up in the shop now and then, especially for putting together woodworking jigs or any piece that may need to be disassembled at some point.

Machine bolts are either threaded all the way to the head or, most often, have an unthreaded shank. They are referred to by their diameter and length, the number of threads per inch, the material they're made of, and the type of head they have. Note that bolts' lengths are taken from the end of the bolt only to the underside of the head, not to the top.

All bolts come in coarse, fine, or extra-fine threads; coarse threads are suitable for most woodshop applications.

Carriage bolts have round tops with a square shank underneath to prevent the bolt from twisting once it's in the wood. Carriage bolts are sold in lengths ranging from 1" to 8".

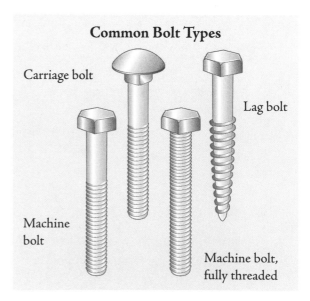

Common Bolt Types

Carriage bolt

Lag bolt

Machine bolt

Machine bolt, fully threaded

Washers, which are made of steel, are used as a protective measure between the head and a project's finished surface. They are sold according to hole diameter, ranging from ¼" to ¹³⁄₁₆".

standards, brackets, and pins

Nails, screws, and bolts hold pieces of wood together permanently (or at least that is the idea). Sometimes you want to hold pieces together in a way where you can remove or adjust their elements later. That's where standards, brackets, clips, and pins come in. These fasteners, usually made of metal, are used mostly for connecting shelving.

Standards come in a variety of materials and colors, such as zinc-plated, white-epoxy-coated, brass-finished, heavy-duty steel, or in a variety of hardwoods. They are slotted and attach to the case's sides. Clips or brackets are inserted into the standards' slots. Brackets can handle heavier loads than clips.

Fixed brackets come in wood or metal. They attach directly to a case or to a wall, and are often more ornamental in nature, in contrast to the rather utilitarian nature of standards.

Shelf pins are usually metal or plastic. Pins are installed in holes that are drilled in the sides of the cabinets. The shelving sits on the pins for support. You can buy (or design your own) jigs to help you space your shelf-pin holes accurately. Shelf pins commonly come in ¼" or 5mm sizes. They can be used to support wooden and glass shelves.

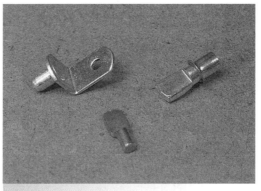

▶Left to right: a ¼" shelf support pin with a bracket; a 5mm pin that will hold a tremendous amount of weight; and a 7mm pin with a collar—the collar is installed in the hole and the pin is inserted, which adds extra support for the pin when using softwoods like pine.

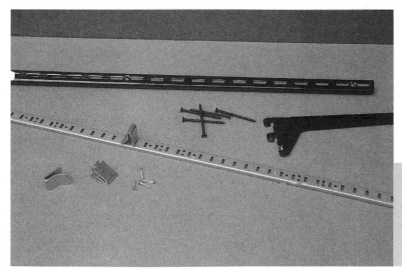

▶Standards are slotted to accept clips (lower left) or brackets (center right). The standards can be mounted flush by setting them into a groove in the cabinet's sides.

> Hinges

When you are joining a door or a lid to a project, you'll need to head down to the hardware store or home center and consider what sort of hinges you need. You may be surprised at the number and variety of choices you'll find.

Hinges fall into one of two general categories: exposed or hidden. *Exposed hinges,* which are a bit easier to install, are very common and can be found at almost any hardware or home-improvement store. *Hidden hinges,* also called Eurohinges, have been, in the past, a little more difficult to find, but they are gaining in popularity.

Exposed hinges are attached with screws to the outside of a face frame. Hidden hinges are attached with screws to the inside edge of a face frame. Both types are attached with screws to the inside face of the door.

While hinges usually fit into one of these two categories, that's where the simplicity ends. There is an incredible array of hinge types, styles, and applications.

butt hinges

Butt hinges are the most common of all furniture hinges; you're also likely to find several in each door frame in your house. Basically, they are two leaves of metal joined together by *barrels,* or *pivots,* that allow them to open and close easily. One leaf is recessed into the cabinet, and the other leaf is recessed into the frame. The two leaves are folded together when the door is closed.

Butt hinges are referred to by their length and width when fully open. So, a 2×1 butt hinge is 2" long and opens fully to 1".

Piano hinges are a type of butt hinge, only longer. They are often available in lengths up to 6' and can be cut to size

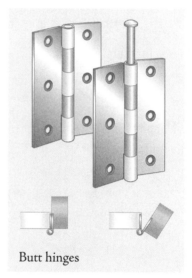

Butt hinges

at most hardware outlets. They are a good choice when strength is needed, such as when attaching a large lid to a blanket chest, because their greater length provides a number of places for attaching screws.

Knife hinges

knife hinges

Intended for applications that don't require an incredibly strong hinge, *knife hinges* consist of two interlocking leaves that turn toward one another. They are inserted into a slot that has been routed into the top and bottom of a door. The hinge is then attached to both the door frame and the door with screws. Knife hinges remain visible from the front or side of the cabinet.

eurohinges

Often referred to as *Eurohinges* or *European hinges,* these hidden hinges are gaining more and more in popularity. They are very easy to adjust, are often used in contemporary cabinetry, and are completely hidden from view after installation.

Other types of hidden hinges are also available, including brass quadrant hinges that are ideal for smaller woodworking projects, like humidors and jewelry cases.

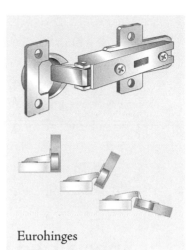

Eurohinges

offset hinges

Offset hinges are commonly used on kitchen cabinets or utility cabinets. To attach the hinge, you first cut a groove (called a *rabbet*) along the inside edge of the door, which creates a lip. The mounting plate of the hinge fits into the lip and wraps around the inside of the door. The other plate is attached to the face of the cabinet. These hinges are available in a ⅜" (as shown in the illustration) or a ¾" offset (which requires no rabbet).

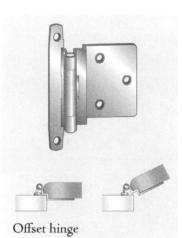

Offset hinge

drop-leaf hinges

Drop-leaf hinges have a special application. One side of the hinge is longer than the other so a ruled joint can be used (see illustration) where the tabletop and leaf come together. This also makes it easier to clean the table and not get crumbs in the barrel of the hinge. These hinges can be attached directly to the bottom of the table without mortising.

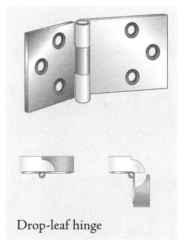

Drop-leaf hinge

swaged hinges

Butt hinges come in two different configurations as to how they close. In the illustration on page 38, the hinge on the right is typical of continuous hinges and most butt hinges for cabinetry. The leaves of the hinges at the barrel are not bent, so the hinge closes with the leaves separated. If this hinge were mounted directly on the cabinet and the door, there would be a gap the thickness of the hinge barrel between the door and the cabinet side. A mortise is needed to set the hinge in so the gap between the door and the cabinet side is less than the thickness of the hinge.

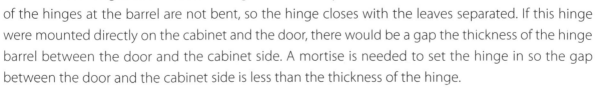

The hinge on the left in the illustration has been swaged, which means one leaf has been bent at the barrel so the leaves come together when the hinge is closed. This can eliminate the need for cutting a mortise for the hinge when it is mounted on the cabinet.

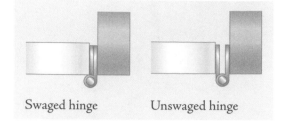

Swaged hinge   Unswaged hinge

The hinges for room doors are usually swaged so the doors can be closed with a small gap at the hinge side to keep the door from binding.

> Specialized Hardware

Depending on the project you're making, you may need more-specialized hardware. Whether you want to take apart a bed frame and put it together, open and shut a drawer, or roll a tool cabinet around your woodshop, you can find hardware designed to do the job.

bed rail brackets

There are two common ways to hold the side rails, footboard, and headboard of a bed together. One is by using *bed bolts,* which are installed by drilling holes into the legs of the bed and inserting the bolts into the ends of the side rails, which have matching holes drilled into their ends. The bolt is screwed into a captured nut in the end of the side rail. The bolt holes in the legs are usually covered with a brass decorative plate that swings open and closed. Beds that are made to resemble antique four-poster frames often use this method.

The other method of putting a bed together is by using a two-part bracket. The female part of the bracket is set into a mortise in the legs, and the male part is set into the ends of the side rails. The male part usually has two hooks that interlock with slots in the female part. This hardware is very strong, and it is the easiest way to put together and take apart a bed.

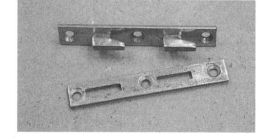

▶Two-part bracket for attaching bed rails

▶Bed bolt attaching a bed rail to a headboard (The dowels in the bed rail are left unglued, and serve to keep the rail in place.)

casters and furniture glides

If you have a heavy cabinet or piece of furniture, sometimes it's good to mount *casters* on the bottom of the piece. Casters come in sizes ranging from 1"- to 5"-diameter wheels. Depending on the floor surface (carpet, concrete, wood, or tile), different types of wheel materials are available that will make moving the piece easier and not harm the floor. Smooth nylon, plastic, or steel wheels are good to use on carpet. If you have a heavy piece of machinery in your woodshop, 3"- to 5"-diameter steel wheels can make it a breeze to move.

Glides are used when a piece of furniture needs to be elevated very slightly off the floor or when a cabinet needs to be leveled.

Furniture glides are available in steel, rubber, and hard nylon. They can be adjustable steel feet with a threaded rod that inserts into a nut on the bottom of the furniture. Rubber and nylon glides come with a nail shank that is simply driven into the bottom of a leg or cabinet.

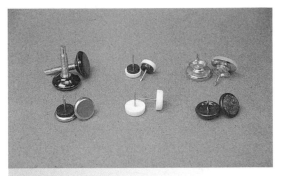

▶Furniture glides come in a huge assortment. Clockwise starting at upper left: adjustable leveling feet; nylon glides with rubber cushions; nylon glides with swivels; nylon glides with felt feet; nylon glides; and metal feet with rubber cushions.

drawer slides

Fitted drawers are exactly that: The drawer is the same size as the cabinet opening and is fitted to that opening using no hardware. But for the rest of us who don't fit drawers, a huge variety of drawer slides is available.

Full-extension drawer slides allow the drawer box to be opened completely and are most commonly mounted on the sides of the drawer box. These slides are rated from light duty to very heavy duty. Heavy-duty slides, for file or tool drawers, have ball-bearing runners that will last for years. Undermount versions of full-extension slides are also available. All side-mount slides require the drawer box to be built 1" smaller than the opening it goes into.

Three-quarter extension drawer slides allow the drawer box to be opened about two-thirds to three-quarters of its depth. These slides are rated light to medium duty and operate very quietly and smoothly. They have indents that prevent the drawer from being pulled out without first lifting slightly on the front of the drawer. These slides are most commonly used in kitchens and desks.

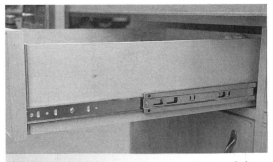

▶Full-extension, side-mounting drawer slides

Undermount drawer slides are a good choice when you don't want to see the hardware. Also, the drawer box can be made almost the same size as the opening it goes into. Two basic undermount styles are available: a single monorail center-mount slide and a two-rail system.

knobs and pulls

After you've built your furniture or cabinetry, it is time to choose the hardware that will help you open and close the doors and drawers. *Knobs* and *pulls* are available in wood, plastic, metal, glass, etc. A complete spectrum of colors and textures is also available.

▶This drawer front has been carved to accentuate the placement of a wooden knob.

The type of hardware you choose for your projects is purely up to you, but here are some basic guidelines: Knobs are generally used on smaller-faced drawers and doors. Pulls are used on larger drawers and doors. Two knobs or pulls on large dresser drawers are acceptable, as are oversize knobs and pulls to achieve a special effect.

As a guide to choosing hardware, consider the type of materials used to make the project, the style, and the size. How will the project be used? Does it match anything else in the room or the house? What color is it?

> Installing Hardware

Installing hardware can sometimes be tricky, but with enough practice and patience, you'll be an expert in no time. The following instructions will give you the know-how to install two common hardware pieces.

installing a butt hinge

The first photo shows a swaged butt hinge (if you need to, turn back to page 36 for a discussion of swaged versus unswaged hinges). The leaves have been peened at the barrel so they come together flat. This makes it possible to install this hinge without cutting a mortise—that is, without cutting a space in the wood to allow the hinge leaf to be flush to the wood's surface. If it were installed without using a mortise, the gap between the door and cabinet side would be the thickness of the two plates when they are closed.

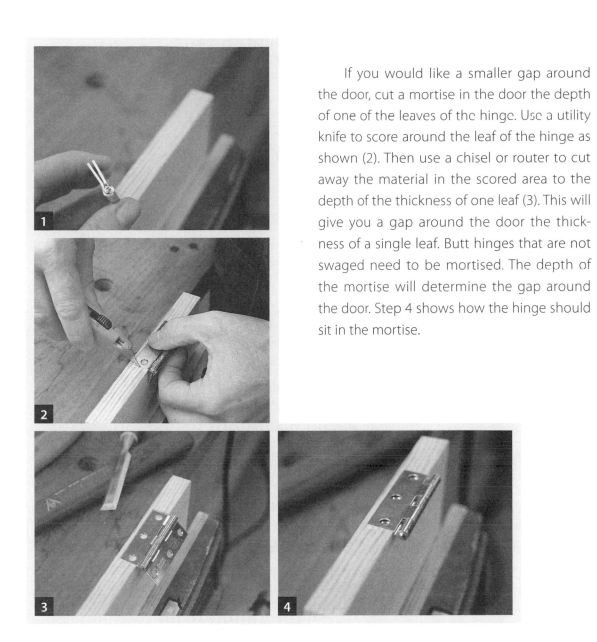

If you would like a smaller gap around the door, cut a mortise in the door the depth of one of the leaves of the hinge. Use a utility knife to score around the leaf of the hinge as shown (2). Then use a chisel or router to cut away the material in the scored area to the depth of the thickness of one leaf (3). This will give you a gap around the door the thickness of a single leaf. Butt hinges that are not swaged need to be mortised. The depth of the mortise will determine the gap around the door. Step 4 shows how the hinge should sit in the mortise.

installing a flush-mounting hinge and catch

Hardware of this type might be used to attach the lid of a blanket chest or a jewelry box. First, position the lid or door on the box or cabinet. Then draw a line that shows the center location of the hinge. Put the hinge in place (1). Drill a hole and install the center screw in the box. Then do the same for the center hole in the lid. Install the second hinge the same way. If the lid works fine and it lines up after being opened and closed a couple of times, install the rest of the mounting screws (2).

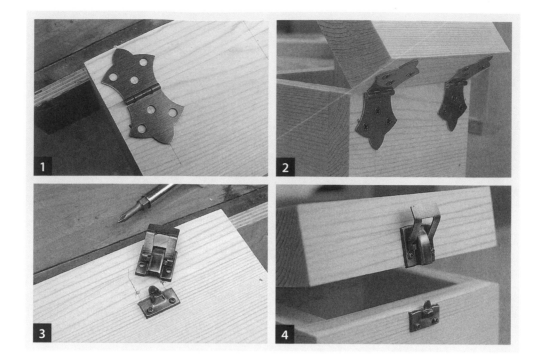

Installing a catch is the same as installing the hinges. Close the lid, and draw a center line or center the catch on the box. Attach the bottom part of the catch (3). Line up the top part of the catch by closing it over the lower part. Then attach the top part (4).

> Glues

If your only experience with glues has been a bottle of white Elmer's school glue (complete with that twist top that is almost always glued tightly shut), you might not be familiar with the wide variety of glues that are out there on the market. Types range from all-purpose yellow glues to contact cement to epoxy glues to polyurethane glues. Each has its own adhesive specialty, and each has its own advantages and disadvantages.

Besides the obvious function of holding two pieces of wood together, adhesives also perform the function of transferring and distributing stress, which increases the strength and firmness of the pieces you build.

adhesives defined

Adhesives generally fall into three basic categories: animal glues, vegetable glues, and synthetic resin glues.

Animal glues are made from collagen, the primary protein that comes from an animal's skin, bone, or muscle, mixed with hot water. Animal glues can also be made from the serum albumin that comes from either fresh blood or a dried-blood powder; these are mostly used in the creation of plywood.

Vegetable glues are often made from the starch and dextrin found in corn, wheat, potatoes, or rice. Gum extracted from trees can also be used as an adhesive.

Synthetic glues are man-made polymers that can be changed or modified to meet particular woodworking needs. They have incredible water resistance and are divided into two basic categories: thermosetting adhesives and thermoplastic adhesives.

activating glues

Adhesives can be activated in different ways. *Thermosetting adhesives* are activated via a condensation reaction in which water is eliminated, putting the adhesive through an irreversible chemical and physical change that makes it insoluble. Chemicals, heat, or both act as the catalyst to force this change. Thermosetting adhesives include urea-formaldehyde, melamine formaldehyde, phenol formaldehyde, and resorcinol formaldehyde.

Thermoplastic adhesives are prepolymerized and do not go through a chemical linking reaction as they cure, which means they remain in a reversible state and can be softened by heating. Thermoplastic adhesives include white glues (polyvinyl acetate emulsions) and hot-melt glues.

glue varieties

Here's a look in more detail at some of the many choices you have when looking for the right glue or adhesive. A chart at the end of this chapter will give you a quick summary of the important differences between the ways that adhesives work.

animal protein glues

Animal protein glues, which include hide glues, come in both solid and liquid forms. Solid forms must be added to water, melted, and kept warm during use; they have low resistance to moisture. These glues are sometimes used in the assembly of general furniture projects, but mostly work well for constructing stringed instruments and repairing antique furniture. Hide glue, specifically, is inexpensive and dries clear, is fairly useful for filling in gaps, and is nontoxic, but it's not waterproof.

blood protein glues

Blood protein glues are solid and are mixed with water and chemicals such as lime or caustic soda and have a little more resistance to moisture than animal protein glues. They're generally used for interior-use softwood plywoods, but they have been replaced in the woodworking world with phenolic adhesives.

▶Left to right: extended exterior wood glue; liquid hide glue; yellow exterior wood glue; white glue; yellow wood glue; and brown wood glue.

white and yellow glues

The most common and popular glues are white and yellow glues.

White glues, known scientifically as polyvinyl acetate, are commonly used for general woodworking, model-making projects, and repairs that involve porous materials such as paper or leather. White glues give a colorless bond line and are applied in liquid form directly to the wood and must be pressed at room temperature. White glues set quickly, but temperatures over 100°F can weaken a joint held together with white glue.

Known scientifically as aliphatic resin glue, *yellow glues* are currently favored by most woodworkers for most woodworking projects. These glues come in liquid form and are easy to use. They are great for a variety of applications, but they don't work well for outdoor-furniture projects or for situations where a water-resistant bond is desired. Yellow glues dry translucent, are nontoxic, inexpensive, and have a slightly heavier consistency than white glues. Not entirely waterproof, yellow glues are more moisture resistant than white glues and less affected by high temperatures.

Brands of white glues include Elmer's Glue-All. Brands of yellow glues include Titebond and Elmer's Carpenter's Glue.

contact cement

Contact cement is used mainly for veneering and for permanently bonding plastic laminates (such as countertops). It bonds instantly on contact and is highly toxic. Because of the fumes, work in a well-ventilated area and wear respiration protection. It's best to apply contact cement with a brush or a roller to both of the surfaces to be bonded. Inexpensive, it bonds clear and is very moisture resistant. DAP Weldwood Contact Cement is a brand of contact cement.

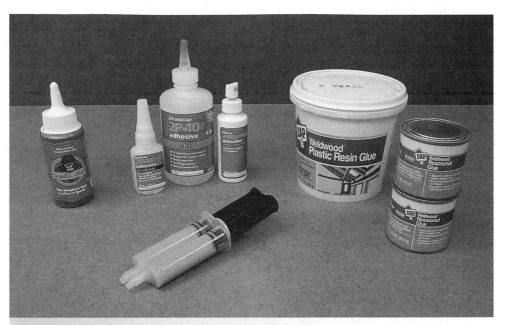

▶Left to right: polyurethane glue; cyanoacrylate and activator (speeds the curing time); dry plastic resin (mixed using water); two-part resorcinol; and in front, two-part epoxy.

plastic resin glue

Known also as urea-formaldehyde glue, *plastic resin* is good for cabinet repairs and for furniture that needs to be incredibly strong. A moisture-resistant adhesive, it is inexpensive and dries opaque. Plastic resin comes in a powder form that must be mixed with water and is toxic until fully cured. It also requires a lengthy clamping time. Brands include DAP Weldwood Plastic Resin Glue.

cyanoacrylate glue

Cyanoacrylate is known generically as superglue and is used for quick, small repairs. It is inexpensive, bonds to almost any material (plastics, metals, vinyl, rubber, ceramics, and wood), and dries incredibly fast. It's fairly resistant to moisture and dries clear. This toxic glue comes in either a liquid or gel form; be very careful not to let it come in contact with your skin. Brands include Krazy Glue and Duro Quick Gel.

resorcinol glue

Incredibly waterproof, *resorcinol* is used not only for outdoor-furniture projects, but also boatbuilding and other marine uses. Expensive, resorcinol dries opaque and is toxic until it is fully cured. It comes in a powder form that must be mixed in a liquid resin and must be used within a few hours of being mixed. Although it's very strong and dependable, resorcinol does leave an obvious glue line. Brands include DAP Weldwood Waterproof Resorcinol Glue.

epoxy glues

Epoxy glues are used to bond unlike materials, such as metals to wood or glass to particleboard. Epoxies are expensive, but they are waterproof and strong. These glues come in two parts, either liquid or putty, that must be mixed right before using and are toxic until fully cured. Brands include Devcon 2-Ton Epoxy.

hot-melt glues

Hot-melt glues are used for quick repairs on leather or fabrics and are also good for making small repairs and filling joint gaps on furniture. Be careful when applying hot-melt glues, as the glue gun used to apply the glue can be very hot to the touch. Certain finishing compounds have been known to affect the strength of hot-melt glues. Brands include Thermogrip Hot Melt.

polyurethane glue

Good for filling gaps and resistant to the consistent presence of moisture, *polyurethane glue* actually cures by being exposed to moisture: Polyurethane changes from a liquid into a foamlike substance when it is applied, expanding out of the joint. Its color varies from clear to brown, but it leaves a fairly colorless bond line. One well-known brand of polyurethane glue is Gorilla Glue.

> Gluing Up Projects

When you're mating joints using a glue or an adhesive, the first thing to do is make sure your wood surface is smooth, flat, and free of any machine-caused roughness or chinks. Dirt or roughness in the wood will interfere with the ability of the glue to adhere to the surface. Be sure your wood is accepting of an adhesive and isn't too oily or dirty.

Your wood should also be dry, as water can dilute the power of the adhesive you're using.

Precision really is important when creating a glued joint. You can't rely totally on the power of the glue. For a successful edge joint, the surfaces that will be mated must match perfectly and tightly. Be sure the glue is spread evenly on the pieces to be joined, but don't overdo it. Less can be more when you use adhesives to build your projects. The more squeeze-out you have, the more scraping and cleaning you'll have to do later.

Plan your glue-ups with an eye on the clock, and with some logic as to the order of the pieces you'll be gluing. Keep in mind that, depending on things like temperature and humidity, every type of glue has a fairly specific amount of time it can be left open and exposed to air before you assemble your pieces, as well as a certain amount of time it takes to set.

Common Adhesives

Adhesive	Advantages	Disadvantages	Common Uses	Working Time	Clamping Time at 70°F	Cure Time	Solvent
Yellow glue (aliphatic resin)	Easy to use; water-resistant; leaves almost invisible glue lines; economical	Not waterproof (don't use on outdoor furniture)	All-purpose wood glue for interior use; stronger bond than white glue	5 to 7 minutes	1 to 2 hours	24 hours	Warm water
Contact cement	Bonds parts immediately	Can't readjust parts after contact; leaves unsightly glue lines	Bonding wood veneer or plastic laminate to substrate	Up to 1 hour	No clamps; parts bond on contact	None	Acetone
Superglue (cyanoacrylate)	Bonds parts quickly	Limited to small parts	Bonding small parts made from a variety of materials	30 seconds	10 to 60 seconds; clamps usually not required	30 minutes to several hours	Acetone
Epoxy glue	Good gap filler; waterproof; fast-setting formulas available; can be used to bond glass to metal or wood	Requires mixing; expensive; difficult to clean up; very toxic	Bonding small parts made from a variety of materials; bent laminations	5 to 60 minutes, depending on epoxy formula	5 minutes to several hours, depending on epoxy formula	3 hours and longer	Lacquer thinner
Animal glue, dry (hide glue)	Extended working time; water cleanup; economical	Must be mixed with water and heated; poor moisture resistance (don't use on outdoor furniture)	Time-consuming assembly work; stronger bond than liquid animal glue; interior use only	30 minutes	2 to 3 hours	24 hours	Warm water
Animal glue, liquid (hide glue)	Easy to use; extended working time; economical	Poor moisture resistance (don't use on outdoor furniture)	Time-consuming assembly work; interior use only	5 minutes	2 hours	24 hours	Warm water
Polyurethane	Fully waterproof; gap-filling	Eye and skin irritant	Multipurpose, interior and exterior applications including wood to wood, ceramic, plastic, solid-surface material, stone, metal	30 minutes	1 to 2 hours	8 hours	Mineral spirits while wet; must abrade or scrape off when dry
White glue (polyvinyl acetate)	Easy to use; economical	Not waterproof (don't use on outdoor furniture)	All-purpose wood glue for interior use; yellow glue has stronger bond	3 to 5 minutes	45 minutes to 2 hours	24 to 48 hours	Warm water and soap
Waterproof glue (resorcinol)	Fully waterproof; extended working time	Requires mixing; dark color shows glue line on most woods; long clamping time	Outdoor furniture, marine applications	20 minutes	1 hour	12 hours	Cool water before hardening
Plastic resin (urea-formaldehyde)	Good water resistance; economical	Requires difficult mixing; long clamping time	Outdoor furniture, cutting boards; good for veneering	15 to 30 minutes	6 hours	24 hours	Warm water and soap before hardening

If your shop is hot and dry, your glue will generally set more quickly. If your shop is too cold, the most common glues, such as yellow glues, often don't work well. Try to create an environment that is not too dry, not too moist, and is a nice, midrange temperature, at least 55°F.

Clamps are an essential accessory to the glue-up process. Don't get too intense with your clamps and put them on too tight. Set them gently but firmly, and wait for the squeeze-out to appear. After it has set for about a half-hour, scrape off the dried squeeze-out. When using regular yellow wood glue, you'll probably want to leave your clamps on for at least an hour, then leave the glued pieces alone overnight for the glue to fully cure.

glue types and tips

When using polyvinyl acetate (white glue), spread the glue evenly and clamp the mating pieces together.

Aliphatic resin (yellow glue), the woodworker's glue of choice, should be spread on evenly and clamped. Because yellow glue has a thicker consistency than white glue, it won't drip and run out of the joint quite as much.

If you're using old-fashioned hide glue, be sure to apply it to both of the surfaces that will be mated and let both pieces sit before putting them together. The glue should be in a somewhat "tacky" state before you try to mate the pieces.

Be in a well-ventilated work area when using contact cement, as you'll experience strong fumes while you coat both surfaces to be mated. Let the contact cement dry before you mate the pieces to be joined. Contact cement bonds almost immediately.

If you don't care about a very visible glue line and you're building something that absolutely must be waterproof, resorcinol is a good choice. But it involves a little work. You'll have to mix the resorcinol powder in liquid resin and use it within a few hours. Once you've done that, coat and clamp as with other glues, but be careful, as resorcinol is toxic until it has fully cured.

If you're bonding unlike materials, such as metal to wood, epoxy works well but does involve some mixing. A good gap-filler, epoxies have to be used within a few minutes of the mixing process.

gluing different types of wood

It helps to understand that, regardless of the type of glue or adhesive you use, certain wood species bond more easily than others.

Hardwoods that bond fairly well, but with some minor difficulty, include: white ash, beech, birch, cherry, hickory, madrona, hard maple, and red and white oak, as well as the imported bubinga species. Softwoods that bond fairly well include yellow cedar and southern pine.

Hardwoods that bond well include butternut, elm, soft maple, sycamore, tupelo, black walnut, and yellow poplar. Softwoods that bond well include Douglas fir, ponderosa pine, and sugar pine. Mahogany, an import, also bonds well.

The wood species that bond most easily include the following:

Hardwoods: alder, aspen, basswood, cottonwood, magnolia, black willow

Softwoods: white and Pacific fir, eastern white and western white pine, western red cedar, redwood, and Sitka spruce

Imported woods: balsa and purpleheart

chapter 4

Sanding and Finishing

For many woodworkers in the middle of a project, a good part of the excitement seems to vanish once the dovetail saw and the chisels, or the router and the band saw, have done their work. At that point, few of us can muster great enthusiasm for the work that lies ahead—sanding, staining, and finishing.

This is understandable. Sanding and finishing don't provide the immediate gratification that comes from seeing pieces of wood change form and join other pieces to become something entirely new. And the process of finishing, particularly sanding, can take some pretty hard work. But if you decide to cut corners so close to the end—at the finish line, as it were—you're only cheating yourself. It may seem like a good idea now to skip that one pass with another grit of sandpaper, or not to rub out the finish yet one more time. But thorough sanding and a good finish are what make the difference between a great piece of work and one that is merely adequate, and between one that will be treasured for years to come and one that will require refinishing before you know it.

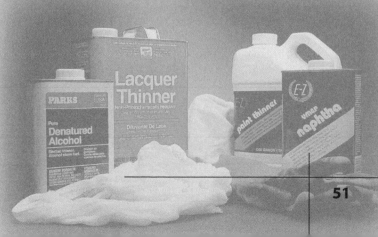

> Sanding Materials

When you're finishing a woodworking project, you start and end with sandpaper. Although the serious sanding takes place before you apply your first coat of finish, you still need to use some elbow grease to buff and shine even after your final coat has been applied.

If you plan to stain the wood, you need to sand it first to prepare the wood surface. When sanding, be sure to remove all wood fibers and open up the wood grain; this will ensure a penetrating, uniform finish.

This preparation process also includes removing any other serious defects you see, such as grease or indentations. If you don't have a clean, smooth surface to work with, the stain and the finish won't sink into the wood pores evenly.

When you reach the staining phase, the penetration of your chosen stain in a hardwood will ultimately depend on the final grit you used to sand the wood. Grits range from 12, which is extremely coarse, to 1,500, which is incredibly fine. You are not likely to use either of those grits, though, as woodworkers generally use from 40 to 600 grit.

In the process of sanding, you will be "working through the grits." This means that you start with a rougher sandpaper (lower grit number) and work your way up through a series of progressively smoother grades (higher grit number) to a final finish. At each step, the smoother sandpaper removes the marks left by the lower grade and leaves some of its own. In general, the progression of sandpaper grits is as follows:

40 to 60—Coarse: Heavy sanding; stripping off paint or other finishes; roughing up the surface

80 to 120—Medium: Smoothing the surface; removing imperfections

150 to 180—Fine: Final sanding pass before finishing the wood

220 to 240—Very Fine: Sanding between coats of stain or sealer

280 to 320—Extra Fine: Removing dust or small marks between finish coats

360 to 600—Super Fine: Fine sanding of the finish

You can use 40- to 50-grit sandpaper for smoothing very rough wood surfaces, though for wood that is reasonably smooth you are better off starting with a medium grit. Before applying any stain or finish, most woodworkers sand using up to a 180 grit, and some use up to 200 or 220 grit. Depending on what type of finish you are applying, the entire finishing process may take you up to a 400 or even a 600 grit.

Don't panic: you don't need to hit every stop along the way in your progression—that would make for truly endless sanding. But for a clear and smooth finish, you will need to employ a minimum of three (and more likely four to six) different grits.

When moving through your grit sequence, try not to skip more than one grit number. This will lengthen the life of your sandpaper and whatever implement you're using to engage it, and it will give you a higher-quality finish.

Look on the back of a piece of sandpaper. There, you'll generally find the product number, lot number, abrasive type, grit size, open or closed coat, and type of backing of the sandpaper. "Open coat" means that only 50 to 70 percent of the sanding surface is covered with particles of grit; "closed coat" means that the entire surface is coated with grit. Sandpaper backing weight is rated by a letter designation. For example, J-weight cloth backing is lightweight; × is of a medium weight; and Y is heaviest.

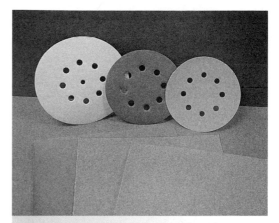

▶The back row is a sampling of sanding disks. In front, from left, sanding papers of these varieties: silicon carbide wet/dry; garnet; and aluminum oxide.

After your wood has been sanded and cleaned, you might want to test your work. Before beginning the coloring process, sponge the surface of the wood with water, alcohol, or any solvent to reveal any areas that might contain glue, marks, or any uneven sanding.

types of sandpaper

Sandpaper comes in several types, which are manufactured for specific uses. These include the following:

Aluminum oxide: This manufactured sandpaper usually comes in an off-brown color and is incredibly abrasive. Woodworkers use it mostly for stripping old paint or varnish or for finishing hardwoods. It's also used for finishing some metals.

Emery sandpaper: Emery is a natural abrasive, black in color, and is most often used for lightly polishing metals.

Garnet sandpaper: Garnet is a natural sandpaper, as well, and while it is not quite as abrasive as manufactured sandpapers, it's made a bit tougher from heat treating. Garnet is popular with woodworkers for finish-sanding fine hardwood furniture projects.

Silicon carbide: This manufactured sandpaper is most often black in color and is the hardest of the commonly used abrasives. It's used for hand-sanding both softwood and hardwood projects.

Zirconia alumina: This manufactured sandpaper is most often used for heavy sanding, such as with a belt sander.

> Dyes, Stains, and Fillers

For coloring, you have two main choices: dyes or pigmented stains.

Dyes are basically a mixture of colorants in mineral spirits, oil, alcohol, or water. A dye will change the hue of your wood without concealing its figure. Dyes penetrate both soft and hard grains. Dye particles have smaller molecular structures than the mineral particles found in stains. This is why dyes appear to be more transparent. Dyes also bind to wood naturally and therefore don't require an additional binder.

Stains are created from a variety of sources, ranging from synthetic materials to organic minerals. Stains consist of finely ground pigment particles that are suspended, or dispersed, in either a water-based or oil-based solvent. After the stain is applied, the solvent evaporates, leaving the color on the wood. Pigments are pretty easy to use and come in a wide variety of colors that can be added to other stains to increase color and/or density.

Stains can be applied in a variety of ways, such as spraying, brushing, or wiping. You can manipulate the depth and final color by changing the length of time the stain is left on the wood's surface and how intensely it is wiped off.

Many woodworkers will color their wood project with a dye, then stain it, in order to avoid covering the grain of the wood with the saturation of a dark color. Dyes seem to stain the grain and the areas between the grain about the same color, whereas pigmented stains seem to fill the grain, leaving the wood surface with a little less color.

fillers, sealers, and glazes

In order to fill unsightly pores and smooth out the surface of your wood, you can use a *wood filler*. They can be applied heavily by spraying or brushing. Wipe off any excess with a rag or a scraper to make it flush with the top of the wood's surface.

You must then seal everything with a *sealer*. Some top-coat finishes are self-sealing, or you may need to apply the sealer separately. Often, vinyl sealers are used to lock in the color and protect the grain.

The sealer performs many functions: It locks in the color, seals the grain, begins the filling process, and gives you a coating you can sand.

Glazes are applied after the sealer. These are transparent stains that are used to even out a light and dark area, as you see the true color of wood only after the sealer has been sanded. Tinted applications of a sealer or top coat, called *toners*, can also be used to intensify color.

> Finishes

When it come to finishes, you have a lot to choose from.

Top coats come in a variety of forms, from shellacs to polyurethanes. Each form has different preparation techniques and characteristics that you should keep in mind when making your choice.

- *Danish oil* is easy to use, but it dries very slowly. Amber in color, Danish oil dries to a satin finish and has low moisture resistance.

- *Lacquer* dries quickly and is clear. It is highly glossy, durable, and resistant to moisture.

- *Polyurethane* is another slow drier and comes in gloss, semigloss, and satin. Colors range from clear to amber. Polyurethane is incredibly durable and moisture resistant.

- *Shellac* is an economical, high-gloss, and quick-drying option with color choices ranging from amber to clear. It comes in liquid form, as do all other top coats. It's affected by water, alcohol, and heat, so it is best for indoor projects.

- *Tung oil* is easy to apply as it requires no mixing and comes ready to use, although it does dry slowly. Satin in appearance, tung oil isn't very resistant to moisture.

- *Varnish* takes much longer to dry than lacquer and comes in gloss, semigloss, or satin. Amber in color, varnish is durable and resistant to moisture.

►Shellac is created by dissolving shellac crystals (right) in alcohol (left).

►Shellac is being brushed onto a wood table.

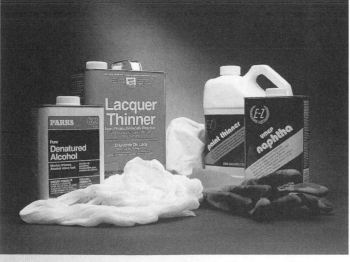

►A selection of thinners

thinning and dissolving finishes

You're probably familiar with the choice you face at your local paint store: using latex paints, which clean up with water; or oil-based paints, which require you to soak your brushes in turpentine or paint thinner. In the paint world, the easier-to-use water-based products have lately been winning the day. When it comes to finishes, water-based products are

Finishing Materials: What Dissolves and Thins What

Substance	Dissolves	Thins
Mineral spirits (paint thinner), naphtha, turpentine	Wax	Wax, oil, varnish, polyurethane
Toluol, xylene	Wax, water-based finish, white and yellow glue	Wax, oil, varnish, polyurethane, conversion varnish
Alcohol	Shellac	Shellac, lacquer
Lacquer thinner	Shellac, lacquer, water-based finish	Lacquer, shellac, catalyzed lacquer
Glycol ether	Shellac, lacquer, water-based finish	Lacquer, water-based finish
Water	–	Water-based finish

becoming more popular, but for most traditional finishing materials you will still need to buy certain substances (that is, not water) to thin and dissolve them. Here's a handy chart (above) that lists a number of materials you can use to dissolve and thin a variety of finishes.

finishing different wood varieties

The Hardwood Information Center *(www.hardwoodinfo.com)* explains that the cellular structure of a wood influences its appearance. Open-grained hardwoods (such as elm, oak, and ash) are all ring-porous species. These ring-porous species have distinct figure and grain patterns. Close-grained hardwoods (such as cherry, maple, birch, and yellow poplar) are known as diffuse-porous species. These species mostly have small, dense pores that result in less-distinct figure and grain patterns.

Note that some close-grained woods (such as cherry and maple) may have a tendency to develop finishing blotches; don't wear yourself out trying to get these blotches out with heavy sanding. You can't get rid of them, no matter how much you sand.

When you get to the staining phase, the penetration of your chosen stain in a hardwood will ultimately depend on the final grit you used to sand the wood.

the durability of finishes, indoors and out

The toughest challenge to a finish is, naturally, exposure to the elements. For outdoor furniture, this may include rain, snow, extreme temperature changes, and more. Even so, the failure of exterior finishes is usually the result of the wrong kind of finish being applied to the wood surface or of not following recommended application procedures.

Each wood variety has unique characteristics that will affect the durability of any finish applied to it.

Dimensional change in lumber occurs as the wood gains or loses moisture. Wood in heated homes tends to dry and shrink in the winter and gain moisture and swell in the warm summer months.

Grain direction affects paint-holding characteristics and is determined at the time lumber is cut. Flat-grained lumber will not hold paint as well since it shrinks and swells more than edge-grained lumber and because wide, dark bands of summerwood are frequently present.

Paint will last longer on smooth, edge-grained surfaces. Penetrating stains or preservative treatments are preferred for rough-sawn lumber. These treatments often accentuate the natural or rustic look of rough-sawn lumber and allow the wood grain and surface texture to show through the finish.

Sanded and rough-sawn plywood will develop surface checks, especially when exposed to moisture and sunlight. These surface checks can lead to early paint failure with oil or alkyd paint systems.

preservatives

Wood preservatives are not considered to be finishes. However, wood properly treated with a preservative can withstand years of exposure to severe decay and insect attack without being affected.

The common wood preservatives are creosote, pentachlorophenol in oil, and the newer waterborne salt treatments—all of which are restricted-use pesticides.

Creosote and pentachlorophenol in oil result in a dark and oily surface. Odor with creosote is a problem. Wood treated with creosote or pentachlorophenol in oil is not recommended for use around the home where people will come in contact with it.

However, wood treated with waterborne salts is suggested for use in patio decks, outside steps, privacy fences, and other home uses. This material is generally light to bright green or brown in color. It can be used outdoors without finishing and will go practically unchanged or weather to a light gray.

▶Polyurethane finish is one of the best for outdoor uses. If you want to make a wiping finish, thin polyurethane with mineral spirits until the finish is a watery consistency. Wipe on and wipe off, then sand between coats. Three coats will yield a highly durable finish.

Common Stains and Top Coats

Stains

Stain Type	Form	Preparation	Characteristics
Pigment Stains			
Oil-based	Liquid	Mix thoroughly	Apply with rag, brush, or spray; resists fading
Water-based	Liquid	Mix thoroughly	Apply with rag, brush, or spray; resists fading; water cleanup
Gel	Gel	Ready to use	Apply with rag; won't raise grain; easy to use; no drips or runs
Water-based gel	Gel	Ready to use	Apply with rag; easy to use; no drips or runs
Japan color	Concentrated liquid	Mix thoroughly	Used for tinting stains, paints, varnish, lacquer
Dye Stains			
Water-based	Powder	Mix with water	Apply with rag, brush, or spray; deep penetrating; best resistance of dye stains; good clarity; raises grain
Oil-based	Powder	Mix with toluol, lacquer, thinner, turpentine, or naphtha	Apply with rag, brush, or spray; penetrating; does not raise grain; dries slowly
Alcohol-based	Powder	Mix with alcohol	Apply with rag, brush, or spray; penetrating; does not raise grain; dries quickly; lap marks sometimes a problem
NGR (non-grain-raising)	Liquid	Mix thoroughly	Apply with rag, brush, or spray (use retarder if wiping or brushing); good clarity; does not raise grain.

Top Coats

Finish Type	Form	Preparation	Characteristics	Dry Time
Shellac	Liquid	Mix thoroughly	Dries quickly; economical; available either clear or amber-colored; high-gloss luster; affected by water, alcohol, and heat	2 hours
Shellac flakes	Dry flakes	Mix with alcohol	Dries quickly; economical (mix only what's needed); color choices from amber to clear; high-gloss luster; affected by water, alcohol, and heat	2 hours
Lacquer	Liquid	Mix with thinner for spraying	Dries quickly; clear (shaded lacquers available); high-gloss luster, but flattening agents available; durable; moisture resistant	30 minutes
Varnish	Liquid	Mix thoroughly	Dries slowly; amber color; gloss, semigloss, and satin lusters; very good durability and moisture resistance; flexible	3 to 6 hours
Polyurethane	Liquid	Mix thoroughly	Dries slowly; clear to amber colors; gloss, semigloss, and satin lusters; excellent durability and moisture resistance; flexible	3 to 6 hours
Water-based polyurethane	Liquid	Mix thoroughly	Dries quickly; clear; won't yellow; gloss and satin lusters; moisture and alcohol resistant; low odor	2 hours
Tung oil	Liquid	Ready to use	Dries slowly; amber color; satin luster; poor moisture resistance; easy to use	20 to 24 hours
Danish oil	Liquid	Mix thoroughly	Dries slowly; amber color; satin luster; poor moisture resistance; easy to use	8 to 10 hours

NOTE: Dry times are based on a temperature of 70°F and 40 percent relative humidity. Lower temperature and/or higher relative humidity can increase drying time.

Woodworking Safely

If you think about it at all, woodworking can be a pretty dangerous activity. You work with razor-sharp blades (which are often whirring at high speeds); you create a lot of sawdust that could find its way into your lungs; and you do all this in an enclosed space that may be filled with ear-splitting noise, or with the chemical fumes from finishing products. Really, how *can* you woodwork safely?

The answer (as with any matter of safety) lies in three things: proper planning, correct procedures, and good habits. Before you cut your first piece of wood, your shop should be set up in the safest way possible, with the right electrical connections, power tool arrangement, and dust-collection system. Once you get going, you will need to *always* follow the right procedures (keep those blade guards on!) and develop good safety habits until they become automatic (the dust mask and ear protection must go on before the router does!). Because the possible consequences of injury in the woodshop are so serious, your commitment to safety needs to be serious as well.

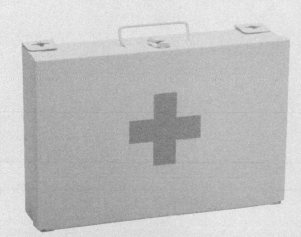

> Shop Safety

Safety is very important when setting up your shop. Read on to learn exactly what you need to do so that you avoid accidents and injuries when woodworking.

basic safety elements in your shop

The most basic safety elements you need are things like goggles (safety glasses), a push stick and push blocks, dust masks, and hearing protectors (earplugs). Have patience: always read your power tool manuals before you start cranking up the decibels and stirring up the dust.

Next, think about paint. Most woodworkers don't stop to think about painting the interior of their shop, but it makes sense to address your walls before you start moving heavy equipment into the room. You don't want moisture in your shop, so it's smart to at least paint your walls with something that will seal up cracks and help the dust to glide off easier.

Then, think about your floor. Be sure your workshop floor space is hard-wearing, void of things to trip and slip on, level, fireproof, and easy to clean and dry. Most woodworkers have home shops with solid concrete slab flooring. If you want additional traction, a cheap solution is to paint that area with a rubber-type adhesive, sprinkle sand on it, and sweep away excess sand after it has dried on the adhesive. Feel free to use rubber mats for traction or in places where you will stand for long periods of time.

Placement of machinery is also crucial. Remember to take into account lighting needs and what your power outlets can handle.

electrical safety

How electricity will be run through your shop can be a simpler problem to solve if you're starting to build a new shop. But if you're like most and already have a shop, as you add a number of bigger and more powerful tools, you will encounter new electrical difficulties.

Start by updating existing wiring if you're turning an existing building into a woodshop or if you've recently added major, new tools to your collection. Then consult the *National Electrical Code* (published by the National Fire Protection Association, or NFPA) for specific guidelines, but be aware that many major cities will have their own specific codes in place. Refer to the codes implemented in your area for specific information and guidelines. Your tool manual will tell you the current capacity of a particular tool's motor. Take this information and match it with the *National Electrical Code*'s wire size recommendations for the expected current usage.

When you've decided where to place major machine tools, you'll need the most power supplied there. That is, you'll need 240-V circuits available for your major woodworking tools (for tools like stationary table saws and planers), as well as a good number of 120-V outlets for your smaller hand-held power tools. Think about your finishing equipment power needs, too.

Place your outlets high up on your walls—chest level or higher—rather than low on the ground. This way you don't have to bend over often and the power cords won't be gathered on the floor where you can trip over them.

Circuit breakers are installed in homes to turn off your electrical system if you're asking too much of it. If you keep tripping your circuit breakers while you're working in the shop, look closely at your wiring setup and your tool placement.

The NFPA recommends replacing or repairing loose or frayed cords on all of your electrical devices, including your woodshop power tools. Some other important tips:

◆ Avoid running cords across doorways or underneath carpets.

◆ Put plastic safety covers over plugs if they are within reach of small children or animals.

◆ Don't overload your outlets.

◆ Always use light bulbs that match a particular lamp's recommended wattage.

In order to protect yourself from shocks and burns associated with power tools, all electric tools should have a three-wire cord with ground and be plugged into a grounded receptacle, be double-insulated, or be powered by a low-voltage isolation transformer.

Three-wire cords contain two current-carrying conductors and a grounding conductor. Any time an adapter is used to accommodate a two-hole receptacle, the adapter wire must be attached to a known ground. Never remove the third prong from the plug!

important fire safety rules

Every woodshop needs fire detection and prevention equipment. You should install smoke or fire detectors, and always keep at least one class ABC fire extinguisher in an easy-access location. And, of course, never, ever, ever throw water onto "live" (that is, plugged in) machinery.

A portable fire extinguisher can put out a small fire, at least until the fire department can arrive. In the unfortunate occurrence of a fire in your woodshop, use your fire extinguisher if the fire is contained in a small area. Call the fire department and be sure everyone has exited the building.

If you actually do have to use that fire extinguisher some day, the NFPA suggests you remember the acronym PASS when handling a fire:

Pull the pin. Hold the extinguisher with the nozzle pointing away from you, and release the locking mechanism.

Aim low. Point the extinguisher at the base of the fire.

Squeeze the lever slowly and evenly.

Sweep the nozzle from side to side.

Keep your fire extinguisher close to the exit. And remember to have a fire alarm installed in your woodshop. Having one in your home isn't enough.

visitors to the shop

Provide visitors, especially children, with safety goggles and make sure they're a safe distance away from machinery. Keep in mind that many machines (for example, portable planers) spit out waste at child's-eye level. If you have children who are old enough to help out in your home workshop, be sure to educate them in the safe use of the machines. When not in the shop, remove start-up keys and lock the workshop. You may want to consider padlocking your machines.

first-aid kits

Storing a first-aid kit in your shop is not simply a good idea; it's a necessity. Whether you want to admit it to yourself or not, there's always going to be an occasional bump or scratch that you'll have to care for. Having bandages and antiseptics on hand is important so that you can handle various medical emergencies fast.

▶An easily accessible, well-stocked first-aid kit is a must in any woodshop.

protect your hearing

We protect all sorts of body parts when woodworking. Protecting your hearing is just as important. Make sure you don't sacrifice your hearing for your love of woodcrafts. A wide variety of ear protection devices—from foam earplugs to earmuffs—are available. Ear protection devices usually come with a Noise Reduction Rating (NRR) that will tell you how many decibels of noise they are intended to eliminate. Find an ear protection method that you are comfortable with, and use it every time you are involved in any noisy woodworking activities.

> Tool Safety

You need to respect your tools and the power they have if you want to be a safe woodworker. Even hand tools can be dangerous. Both hand and power tools can cause woodshop objects to fall to the ground or into containers full of chemicals, can cause objects to fly through the air, can create dust or fumes, or can simply throw wood chunks.

power tool and electrical safety tips

To prevent hazards associated with the use of power tools, workers should observe the following general precautions from the Occupational Safety and Health Administration (OSHA), a division of the U.S. Department of Labor. (*Note:* If you work in a commercial woodshop and see violations of safety codes, OSHA has a toll-free, nationwide hotline that may be used to report workplace accidents or situations posing imminent danger to workers. The number is 800-321-OSHA.)

◆ Never carry a tool by the cord or hose.

◆ Never yank the cord or the hose to disconnect it from the receptacle.

◆ Keep cords and hoses away from heat, oil, and sharp edges.

◆ Disconnect tools when not using them, before servicing and cleaning them, and when changing accessories such as blades, bits, and cutters.

◆ Keep all people not involved with the work at a safe distance from the work area.

◆ Secure the work with clamps or a vise, which will free both of your hands to operate the tool.

◆ Avoid accidental starting of a shop tool by keeping your fingers away from the switch button while you're carrying a plugged-in tool.

◆ Maintain your tools with care. Keep them sharp and clean so that tools perform the way they're supposed to.

◆ Follow the instructions in the user's manual that comes with each new tool for the lubricating requirements and for changing accessories properly and safely.

◆ Be sure to keep good footing and maintain good balance when operating power tools.

◆ Wear the proper clothing and apparel for working in the woodshop. Loose clothing, ties, or jewelry can easily get caught in moving tool parts.

◆ Remove all of your damaged portable electrical tools from your shop.

Double-insulated tools are available that provide protection against electrical shock without third-wire grounding. Double-insulated tools have an internal layer of insulation that completely isolates the tool's external housing. If you need to use a temporary power source, such as during outdoor construction projects, always use a ground fault circuit interrupter (GFCI).

power switches

OSHA states that certain power tools be equipped with a constant-pressure switch or control, including the following: drills; fastener drivers; disk sanders with disks greater than 2"; belt sanders; reciprocating saws, saber saws, scroll saws, jigsaws with blade shanks greater than ¼" wide; and horizontal, vertical, and angle grinders with wheels more than 2" in diameter.

Certain hand-held power tools must also be equipped with either a positive on-off switch, a constant-pressure switch, or a lock-on control, including the following: disk sanders with disks 2" or less in diameter; grinders with wheels 2" or less in diameter; platen sanders; routers, planners, laminate trimmers, shears, scroll saws, and jigsaws with blade shanks of ¼" or less in diameter.

Other hand-held power tools, such as circular saws having a blade diameter greater than 2", chain saws, and percussion tools with no way to hold accessories securely, must be equipped with a constant-pressure switch that will shut off the power when the pressure is released.

what you should know about hand tools

The hand tools category includes any tool that is powered manually, rather than electrically or some other indirect way. Hand tools include such things as screwdrivers, wrenches, axes, and hand planes.

According to OSHA, the following can occur when using hand tools:

◆ If a chisel is used as a screwdriver (or vice versa), the tip of the chisel may break and fly off, hitting the user or others standing nearby.

◆ If a wooden handle on a tool, such as a hammer or ax, is loose, splintered, or cracked, the head of the tool may fly off and strike the user or others standing nearby.

◆ If the jaws of a wrench are sprung, the wrench might split.

◆ If impact tools have mushroomed heads, such as on chisels, wedges, or drift pins, the heads may shatter on impact, which could send sharp fragments flying toward the user or others.

Always direct your tools away from other people, such as friends and family who might want to be in your shop while you are working. Keep in mind that dull hand tools can be more dangerous than sharp hand tools. So keep your tools very sharp and in good condition and you'll be a much safer woodworker.

When working with hand tools, remember that placing these tools next to flammable substances is not a good idea. When in use, iron or steel hand tools can produce sparks that most certainly could be a source of ignition around any flammable material. There are spark-resistant tools available that are made of nonferrous materials.

what you should know about power tools

The main types of power tools are electric, pneumatic, liquid fuel, and hydraulic. Most home woodworkers need to concern themselves only with the first two on this list.

Always use your tool guards, no matter what and no matter how inconvenient they may seem to you. They're there for a reason.

Any exposed, moving parts of your tool need to be guarded. This includes belts, gears, shafts, pulleys, sprockets, spindles, drums, flywheels, chains, or any other moving parts. Never remove your safety guards while the tool is in operation. Read the manuals that come with the tools. Electric tools can cause electrical burns and shocks, which can lead to serious surface injuries or even heart failure. To avoid this and other injuries, always operate your power tools wearing appropriate footwear; store your electric tools in a dry place when you're not using them; keep your work area well lighted when working so you can see properly (this goes for when you're working with hand tools too); and be sure tool cords are out of walkways so that you don't trip over them.

Pneumatic tools, such as drills, hammers, and sanders, are powered by compressed air. Always wear eye, head, and face protection when using pneumatic tools. First, check to see that a pneumatic tool is securely fastened to the air hose so the two will not become disconnected. An added safeguard is a short wire or positive locking device that attaches the air hose to the tool.

Safety clips or retainers should be installed on pneumatic tools to prevent attachments from being ejected while the tool is in use. Pneumatic tools that shoot nails, rivets, staples, or other fasteners, and that operate at a pressure of more than 100 pounds per square inch, have to be equipped with a special device that keeps the fasteners from being ejected unless the muzzle is being pressed against the work surface.

Just like with real guns, never point a compressed air gun at anyone. As you probably already know, pneumatic tools are often very noisy. Remember always to wear hearing protection.

▶The safety guard that accompanies the table saw is an important safety feature to use, and using a push stick to push wood past its blade is a frequently essential safety practice.

trusty tips on safety

Safe enough yet? Not necessarily. Here are a few more rules to follow for safely using tools in your shop. Before you do anything, consult the tool's operation manual.

◆ Read and fully comprehend all of the warnings and instructions that come packaged with your tools.

◆ Keep your tools in good working condition with regular maintenance and attention. Don't use damaged tools.

◆ Be sure your guards and antikickback devices are in good working order and in their correct positions. Before using a blade or a cutter, check to make sure it is sharp and clean.

◆ Use the proper tool for the proper job.

◆ Double-check your wood for loose knots, nails, and other hazards. These can cause injury and damage your equipment if you aren't aware of their locations in the wood.

◆ Always wear goggles, safety glasses, or a face mask when using power or hand tools. When sanding, wear a dust mask too.

◆ Never wear neckties, work gloves, bracelets, wristwatches, or loose clothing. If you have long hair, wear a hat or tie it back.

◆ Don't forget to wear hearing protection when using loud tools.

◆ Be sure to have a push stick or push block within easy reach before starting any cutting or machining operation. Don't put yourself in awkward positions in which a sudden slip could make your hand hit the blade or cutter.

> Dust Collection and Ventilation

Since wood dust is a natural substance, people are often surprised to learn that wood dust is an incredibly dangerous, not to mention irritating, substance.

It can make you cough, make your eyes tear up, and even make you feel a little nauseous if you breathe in too much or the wrong type. These minor things mostly cause the woodworker short-term irritation, but wood dust can also be a serious health concern. It's possible that it can cause respiratory problems and even cancer. And aside from the dust that can accumulate in your body, wood dust can also be the cause of fire and explosions.

So take your dust-collection accessories very seriously. They're not frivolous or extraneous additions to your shop; good dust collection is a necessity.

Wood dust is produced pretty much anytime you're doing anything with wood in your shop: It's produced from chipping, sawing, turning, drilling, sanding, planing, and jointing.

Sanding is the worst perpetrator because the particles produced are so small that masks often don't filter everything, allowing tiny dust particles to enter the woodworker's unsuspecting nasal cavities, sinuses, and lungs.

And although hardwoods such as beech, oak, and mahogany have been documented to be more dangerous (they produce very fine particles of dust and have links to certain nasal cancers), one of the most hazardous woods is the softwood known as western red cedar. Other softwoods, such as pine, pose less of a risk. Medium-density fiberboard (MDF) is unhealthy for humans because of the bonding agent used, and should be regarded the same as a hardwood in terms of dust danger.

The best defense against wood dust in your body is a properly designed and maintained exhaust system to collect the dust that is produced in your shop. And you should use dust respirators, especially if, for some reason, you do not have a good exhaust system installed.

Concentrations of small dust particles in the air can form a mixture that will explode if ignited, and wood dust burns readily if ignited. Fires can start because of a variety of reasons, such as badly maintained heating units, over-heated electric motors, electric sparks, or cigarettes. Just one more reason to stop smoking, especially when you're working wood.

Slightly less dramatic, but just as serious, is the possibility of slipping or tripping from an accumulation of wood dust on the floor. So keep your floors clear.

A shop dust-collection system

controlling dust accumulation

Always have on hand personal protective equipment, such as goggles or other eyewear, overalls, and gloves. Although laundry isn't often a woodworker's favorite task, you need to wash your woodworking clothing regularly to prevent heavy dust accumulation.

To control your exposure to airborne dust in your woodshop, do the following: Try to use a process that reduces the generation of dust in the first place; have local exhaust ventilation attached to all woodworking machines to prevent dust from entering your shop altogether; and, of course, be sure the equipment you have is in proper working order and always maintained.

When you go shopping for your dust-collection units, know that you'll be looking at five different categories: shop vacuums; single-stage, two-stage, and cyclone dust collectors; and air cleaners that are installed near the ceiling to filter and trap dust particles.

A great reference for comparing and contrasting different dust collectors available is *The Insider's Guide to Buying Tools*, published by Popular Woodworking Books. It describes the features, prices, and uses of many different types of dust collectors.

chapter 6

Joinery

Woodworking is really all about the joinery. If you just wanted to make something out of one single piece of wood, you'd be a sculptor, wouldn't you? It's when two or more pieces of wood need to come together—and stay together for decades—that things get tricky, as well as quite interesting.

Your development as a woodworker could largely be described as the story of how you master (or at least try to master) one joinery technique after another. In this chapter we'll discuss a number of methods of joining wood, from butt joints and dadoes, to mortise-and-tenon joints and dovetails. The techniques you use in any one project will be decided by many factors: your skill level; the amount of time you have; the requirements of the piece you are making; the tools you own; your preference for hand tool versus machine woodworking; and more. In the end, what matters is that the joinery you choose serves to create the best piece possible, and, sometimes, to further your education as a woodworker.

> Your Joinery Options

Take two 2 × 4s. Lay them down on their narrower sides, with one end of the first 2 × 4 against the middle of the second. Drive two nails all the way through the side of the second 2 × 4 so that they then go into the end of the first, and the two pieces are connected in a T-shape.

There you go—you've created a *butt joint*. (This was just a mental exercise, by the way, so we hope that you don't now actually have a large wooden T on your shop floor.) The butt joint has been around pretty much since woodworking began, and it is the simplest method of joining two pieces of wood together. It's also the weakest, for a number of reasons:

◆ In most cases, the only connection between the pieces of wood in a butt joint comes from either glue or from metal fasteners (sometimes both), and not from parts of the wood itself, as in many other joints.

◆ The butt joint usually provides fewer matching wood surfaces for glue to take hold than other joints do.

◆ Depending on how the pieces are aligned, butt joints can easily fall victim to the problems caused by wood movement.

◆ A butt joint doesn't provide any sort of groove or edge to hold one of the pieces in its correct place, as, for example, a dado joint does.

Although better joints are in use, the butt joint is still found in woodworking because of its simplicity. In carpentry, where large, strong nails or screws are used to join pieces together, and small gaps caused by wood movement are not much of a concern, butt joints are more the rule than the exception. In finer woodworking, other joints are more common.

Several types of joints are created by removing wood from one board so that it has a space to hold the end of another piece of wood in place. The wood is usually removed to no more than about one-third the depth of the first piece (and preferably less) in order to be sure the piece maintains sufficient strength. When the cut is made to the middle of a piece, the result is a *dado joint*. When it is made to the end, you have a *rabbet joint*. Elements of these two joints can be combined in various ways, such as in a rabbet-and-dado joint (as shown in the section on rabbets on page 77).

Dado joints are particularly useful for situations such as attaching shelves to the sides of a bookcase. One common use of rabbets is to attach a panel to the back of a cabinet in a way that prevents the panel from being visible from the side.

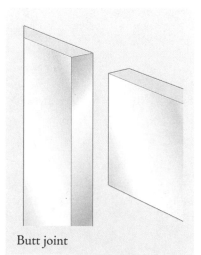

Butt joint

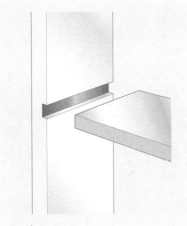

Dado joint

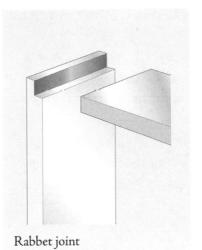

Rabbet joint

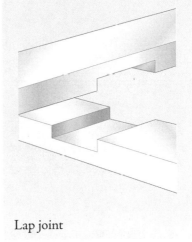

Lap joint

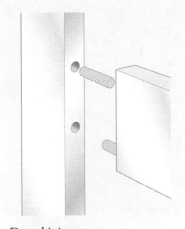

Dowel joint

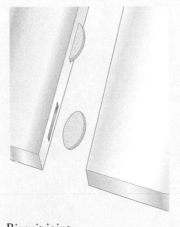

Biscuit joint

When stock is removed from both members of a joint in a manner similar to a dado or rabbet, you have a *lap joint*. If you want the pieces to meet so that they are flush on both sides (as in the illustration above), you remove exactly half the depth from each piece.

One type of joint that is often found in mass-produced, less expensive furniture is the *dowel joint*. In this joint, matching holes are drilled into each piece to be joined, and they are connected by glued-up short wooden dowels. Lining up the holes properly and drilling them straight is best done with a special jig. The main problem with this joint is that for pieces of furniture that get a lot of stress—such as tables, and especially chairs—over time the dowels can work their way loose from either end. This can lead to an unhappy result, especially if you happen to be sitting in the chair when one of its elements finally breaks free.

A more modern joint that has some of the characteristics of the dowel joint is the *biscuit joint*. Here, a special tool called a biscuit cutter uses a small circular blade to cut slots for inserting flat, oval-shaped biscuits. The biscuit cutter itself is used to align the position of the slots properly.

Biscuits create a stronger, more secure joint than dowels for several reasons. They provide much more gluing surface area for greater holding power; and the biscuits themselves expand due to the moisture of the glue, which helps to keep the joint from coming apart down the road.

An evolution of the simple butt joint occurred with the invention of the *mortise-and-tenon joint,* which gives you a very strong, durable joint. According to woodworking historian Gary Halstead, the mortise-and-tenon technique has been used on furniture pieces dating from the late eleventh century.

One common use of mortise-and-tenon joints is to connect the aprons of a table (those long pieces under the tabletop) into its legs. (The ends of the aprons are cut to form a tenon, and mortises are cut into two sides of each table leg.) Other variations of this joint are used in frame-and-panel construction, such as for a cabinet door. One thing to know about mortise-and-tenon joints: If you leave the mortise slightly longer than the tenon, you give the wood in the tenon end room to expand and contract as it moves.

Bridle joints, considered a type of mortise-and-tenon joint, are visually pleasing and offer a large surface area that makes them easy to glue up. Two pieces of wood are interlocked in a T-shaped connection that is very strong.

Dovetail joints are made with a series of fingers that are interlocked for a strong joint. They are useful when you need to join two pieces of wood that are much larger and wider than the narrow wood pieces used for making mortise-and-tenon joints. In traditional cabinetry, the top of a chest of drawers would be attached to the sides with a long dovetail joint, which would usually be covered by molding. Advanced dovetail-makers are able to create concealed dovetail joints.

Dovetails can be cut by hand (which is one sign of a fine piece of furniture) or with a special jig and a router. They are most often used in drawer construction because they are, in one of their directions, essentially unbreakable, so they can resist the front-to-back stresses that are repeatedly placed on a drawer front. A bond of glue is used to keep the joint from opening to the side, where there is much less stress. You might also use dovetail joints to create a blanket chest; in that case you'd probably leave the dovetails proudly exposed as an example of your work.

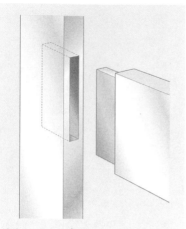

Mortise-and-tenon joint

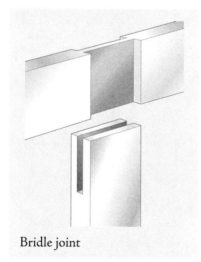

Bridle joint

Dovetail joint

choosing the right joint

When deciding which joint to use in a particular project, you will naturally first think of which joints you feel confident about making properly. As you progress as a woodworker, though, other considerations will come into play. What aesthetics do you desire from your furniture piece? What kind of feel or style are you going for?

Sometimes the style of your planned piece will dictate the type of joint to be used, simply because a particular joint is indicated by a particular furniture style. The mortise-and-tenon joint, for example, is a hallmark of furniture from the Arts and Crafts Movement. Other times, that won't matter, and it will be up to your own tastes.

Some woodworkers just plunge in with their biscuit joiner and make everything with biscuit joints, which is one type of joint construction that is entirely concealed and is based on practicality and not prettiness. Other woodworkers like to implement more finesse and intricacy into their work by using exposed and aesthetically pleasing joints.

Often, though, you'll find yourself using a variety of joint types in a single woodworking project.

Most joints can be made with hand tools, if you consider yourself a craftsman and/or have a lot of patience and respect for the art of woodworking. Most joints, though, can be made with power tools as well, which seems to be the modern method of choice.

Besides aesthetics, the strength you need from a joint is also necessary to consider.

As we've seen, wood is a living substance made of cells that absorb and release water. It moves, expanding and contracting due to moisture gain and loss. This simple fact is what makes joint construction a tricky operation, and it is also the reason the simplest joints to create are sometimes the least strong. In the sections that follow, we'll further explain some of the woodworking joints you can choose from and demonstrate some of the ways they are created.

> Dado Joints

There is an old woodworking saying: "Use the simplest joint that will work." The dado joint certainly works, as it has been used for centuries in cabinetmaking. And it definitely is simple.

A dado is a flat-bottomed channel cut across the grain of the wood. (When it runs with the grain, the channel is called a groove.) You cut a dado or groove into one board, and the mating board fits into it. One well-placed, properly sized cut makes the joint. With today's power tools, it's a cut that is easy to make, if you know how to do it correctly. The first step is to design the joint properly.

The dado doesn't have to be deep to create a strong joint. Usually ⅛" is deep enough in solid wood, while ¼" works in medium-density fiberboard (MDF), plywood, or particleboard. The shallow channel helps align parts during assembly, and the ledge it creates is enough to support the weight of a shelf and everything on it. The dado also prevents the shelf from cupping.

The one stress it doesn't resist effectively is tension, meaning it doesn't prevent the shelf from pulling out of the side. Only glue or fasteners can do that. Because all of the gluing surfaces involve end grain, glue strength is limited.

types of dadoes

When the dado extends from edge to edge, it's called a *through dado*. It's easy to cut. The most common objection to it is that it shows in the final product. However, you can conceal it using a face frame or trim.

Also, the joint can begin at one edge and end before it reaches the other (stopped), or it can begin and end shy of both edges (blind). These dado versions can be a little trickier to cut.

To make a *stopped dado* or *blind dado,* the corners of the mating board must be notched, creating a projection that fits in the dado. Sizing the notches so you have a little play from end-to-end makes it easier to align the edges of the parts. But it does sacrifice a bit of the strength that the narrow shoulder imparts.

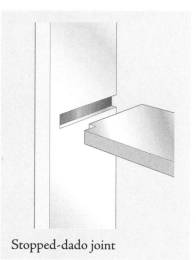

Stopped-dado joint

cutting dadoes

To end up with a strong dado joint, you need to make it the correct width. The bottom needs to be smooth and flat, the sides perpendicular. If the cut is too wide, glue isn't going to compensate; the joint will be weak. You'll need to take care to get the fit right.

The two most obvious power tools for cutting dadoes are the table saw and the router. But there are other options, such as a radial-arm saw. Fitted with a dado head, the radial-arm saw works through dadoes quickly. The work piece is face up, so you can see what you're doing. Layout marks are visible, and you can line up each cut quickly. When a stopped dado is needed, you can cut to a specific mark. The work isn't moved during the cut, so the piece is less likely to twist or shift out of position. This is especially helpful on angled cuts, whether it has a miter or a bevel (or both).

cutting dadoes on the table saw

A table saw is powerful and equipped with accessories—such as a rip fence and a miter gauge—that are useful in positioning cuts for a dado. To use the table saw effectively for dadoing, you need a dado cutter, which is either a stacked set of table saw blades (which is preferable) or a wobble dado set. This is especially true if you have a number of dadoes to cut. If you are cutting only one or two dadoes, then making a number of passes (perhaps five to seven) with any type of blade until the dado is the required width will work fine and not take too much time.

Dado stack sets, which give the cleanest cut, consist of separate blades and chippers. You have to select the combination needed to produce the approximate width of cut desired. To tune the cut to a precise width, you

▶A dado stack set consists of separate blades, washer-like shims, and chippers. You fit the elements onto the saw's arbor, one by one, as shown here.

insert shims between the blades. There's a lot of trial and error in the setup, making test cuts, disassembling the stack, and adding shims.

How do you locate and guide the cut? You may want to use the table saw's rip fence, because it allows you to locate a cut consistently on both sides of a cabinet or bookcase. But a rip fence isn't intended to be used as a crosscutting guide, and dadoes are crosscuts.

If you are going to be making a lot of dadoes, then it will be worth your while to make a jig called a *cutoff box*. It's built specifically for right-angle cuts and rides in both miter-gauge slots (instead of just one). In addition, it effectively immobilizes the workpiece, because the box is what moves, carrying the stationary work piece with it. The work doesn't squirm or twist as you push it into the cutter. Fit the box with a stop so you can accurately and consistently locate a cut on multiple pieces without individual layouts.

▶An accurate, shop-made cutoff box, as shown here, is the best guide accessory to use for dadoing on the table saw.

Stopped cuts (for making the stopped dado shown in the illustration on page 74) can be problematic. Because the work conceals the cutter, and because the cutoff box conceals most of the saw table, it's tricky to determine where to stop the cut. One good option is to clamp a stick to the out-feed table that stops your cutoff box at just the right spot.

A blind cut—used to create a blind dado, which is basically stopped at both ends—would require you to drop the work onto the spinning dado stack. A router may be a better and safer tool for such a dado.

cutting dadoes with routers

The router is often touted as the most versatile tool in the shop, and it certainly is useful for dadoing. The cutters offer convenient sizing: Want a ½"-wide dado? Use a ½" bit. Want a dado for ¾" plywood, which is typically too thin? Use a ²³⁄₂₃" bit, to be sure the joint will fit tightly. Changing bits is quick and easy.

If you have your router mounted in a router table, dadoing with it is much like when you use a table saw. But the router gives you the option of moving the tool on a stationary work piece, and in many situations, this turns out to be the better approach, particularly for dadoing large work pieces, such as sides for a tall bookcase or base cabinet.

To guide the router for the cut, you will need a fence. A shop-made T-square fits the bill, as does a manufactured straightedge clamp, such as the Tru-Grip. An accurate T-square doesn't need to be "squared" on the work, as a Tru-Grip-type clamp does, but positioning it accurately can be tricky.

A setup gauge is helpful here. Cut a scrap to match the distance between the edge of the router base plate and the near cutting edge of the bit. Align one edge of the scrap on the edge of the desired cut that is nearest to where the fence or other guide will be. Position the T-square (or other guide) against the opposite edge and clamp it in place. The guide is set. Holding the router base firmly against the fence as you make a pass will put the dado in exactly the right spot. One word of caution: Depending on the wood you are cutting, the final depth of your dado, and the type of router bit you have, you may need to make several router passes, with a little more depth each time, to make the dado.

▶A crossbar attached at right angles to a plywood straightedge makes it an easy-to-align T-square guide for dadoing with a router. Be sure to clamp it securely to the work and the benchtop at each end.

> Rabbet Joints

The rabbet joint is one commonly tackled by beginning woodworkers. It's easy to cut, it helps align the parts during assembly, and it provides more of a mechanical connection than a butt joint.

The most common form of this joint is the *single-rabbet joint*. In this setup, only one of the mating parts is rabbeted. The cut is proportioned so its width matches the thickness of its mating board, yielding a flush fit.

The depth of the rabbet for this joint should be one-half to two-thirds its width. When assembled, the rabbet conceals the end grain of the mating board. The deeper the rabbet, the less end grain that will be exposed in the assembled joint.

Another common form is the *double-rabbet joint*, in which both pieces are rabbeted. The rabbets don't have to be the same size, but typically are.

ends and edges

The rabbet works as a case joint and as an edge joint. *Case joints* generally involve end grain (that is, where the ends of two boards are joined together), while *edge joints* involve only the sides of boards, where there is long grain. In cases, you will often see rabbets used where the top and/or bottom join the sides (end-grain to end-grain), and where the back joins the assembled case (both end-to-end and end-to-long). In drawers, it's often used to join the front piece with the side pieces.

Because end grain glues poorly, rabbet joints that involve end grain are usually fastened, either with brads, finish nails, or screws concealed under plugs.

rabbet-and-dado joint

One important variant is the *rabbet-and-dado joint*. This is a good rack-resistant joint that assembles easily because both boards are interlocked. The dado or groove doesn't have to be big; often it's a single saw cut, no deeper than one-third the board's thickness. Into it fits an offset tongue created on the mating board by the rabbet.

The rabbet-and-dado joint is a good choice for plywood cases because it's often difficult to scale a dado or groove to the inexact thickness of plywood. It's far easier to customize the width of a rabbet. To make it, cut a stock-width dado, then cut the mating rabbet to a custom dimension. An extra cutting operation is required, but the big benefit is a tight joint.

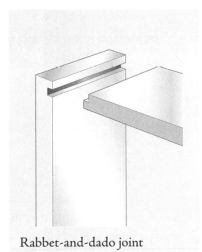

Rabbet-and-dado joint

There are lots of good ways to cut rabbets. The table saw, radial-arm saw, jointer, and router are all tools you might use. The most versatile techniques use the table saw and the router.

rabbets on the table saw

Rabbets can be cut at least two different ways on the table saw. The quickest method is to cut the rabbets using whatever blade is in the saw. Two passes are all it takes. The first cut forms the shoulder. To set it up, adjust the blade height for the depth of the rabbet. There are a variety of setup tools you can use here, but it's always a good idea to make a test cut so you can measure the actual depth of the cut. Then, position the fence to establish the joint's depth by measuring from the face of the fence to the outside of the blade.

To cut the joint, you will first lay the work flat on the saw's table, then run the edge along the fence to make the shoulder cut. If you are rabbeting the long edge of a board, use just the fence as the guide. When cutting a rabbet across the end of a piece, guide the work with your miter gauge and use the fence simply as a positioning device. It is easy to set up, and the miter gauge keeps the work from "walking" as it slides along the table saw's fence. Because no waste will be cut free between the blade and the fence, you can cut rabbets this way rather safely.

Nevertheless, if you feel uneasy about using the miter gauge and fence together, use a standoff block. Clamp a scrap to the fence near the front edge of the saw's table. Lay the work in the miter gauge and slide it against the block. As you make the cut, the work is clear of the fence by the thickness of the block.

Having cut the shoulders of all the rabbets, you next adjust the setup to make the bottom cut. You may need to change the height of the blade or the fence position, or maybe both.

First, adjust the blade to match the width of the rabbet. Reposition the fence to cut the bottom of the rabbet, with the waste falling to the outside of the blade. Make that cut with the work piece standing on edge, its kerfed face away from the fence of the table saw.

When the work pieces are so large as to be cumbersome when held up on their side, you want to cut the rabbets with a dado stack. That way you can keep the work flat on the saw's table. Control the cut using a cutoff box (a sled that runs in the miter slots) or the fence. It's very easy to set the width of the cut with this approach.

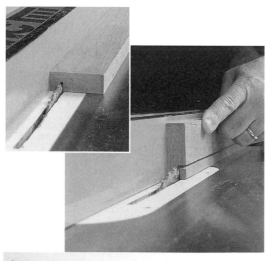

▶ You can saw a rabbet on the table saw in two steps. First, set the blade elevation and fence position to cut the shoulder (left photo). Then adjust either or both to cut the bottom.

Where the proportions of the work piece allow it, use the rip fence to guide the cut. Clamp a sacrificial facing to the fence, as it will be cut by the dado stack. Don't fret about the width of the stack, as long as it exceeds the width of the rabbet you want. Part of the cutter will be buried in the fence facing, and you just set the fence to expose the width of the cutter that's working. Guide the work along the fence.

rabbets with a router

For some jobs, you just want to immobilize the workpiece to your bench and move the tool over it. In those situations, the router is the best choice. A major benefit of the hand-held router approach is that you can see the cut as it is formed. On the table saw (and the router table), the work will usually conceal the cut.

A rabbeting bit is the most commonly used cutter. The rabbeting bit is piloted, which means that it has a ring just below the cutting blade that runs against the part of the wood that is not being cut. This means that you won't need a fence or guide to keep the cut straight and the right width. It typically makes a ⅜"-wide cut. Most manufacturers sell rabbeting sets, which bundle a stack of bearings with the cutter. Want to reduce the cut width? Simply switch to a larger bearing.

If you use an edge guide to control the cut, you can also use a straight bit or a planer bit. Keep the guide in contact with the work piece's edge through the feed—beginning before the cut actually starts and continuing until the bit is clear of the work—and you won't run into trouble.

The edge guide is a big help in beginning and ending stopped or blind cuts as well. Brace the tip of the guide against the work piece edge, shift the router as necessary to align the bit for the start of the cut, then pivot the router into the cut.

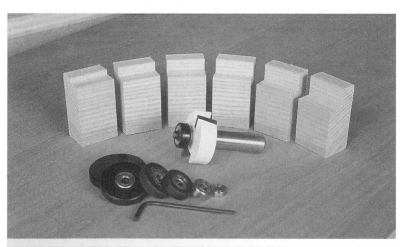

▶One router bit with a selection of bearings enables you to cut rabbets of many different widths.

> Lap Joints

Lap joints are strong, versatile, and easy to cut. You simply cut recesses in both mating pieces, then nest them together. When the recess is half the depth of each piece, they are called half-lap joints, and that is what we will focus on here.

Half-laps can be used for all sorts of flat frames, such as doors, face frames, and picture frames. The half-lap can be used in post-and-rail constructions to join rails or aprons to legs. You usually see this joint in worktables rather than fine furniture, but even in the most traditional table construction, the half-lap is used where stretchers cross (this type of joint is called a cross-lap).

From a practical perspective, the half-lap enjoys an advantage over the mortise-and-tenon joint in that you often need only one tool setup for both parts of the joint. You can join parts at angles quite easily.

Despite its simplicity, a half-lap joint is strong if properly made. The shoulders resist twisting and there is plenty of gluing surface. But be wary of using half-lap joints on wide boards. Wood movement because of environmental changes can break the joint, so confine the joinery to members no more than 3½" wide.

▶ To cut end laps using your table saw, set the height of the blade to half the stock thickness when you want to cut the half-lap shoulders.

cutting lap joints

You can cut half-laps using several different power tools. You can cut them with a router bit using a straight bit; and for the best results, use a sled constructed to hold the piece you are cutting in position.

Be mindful of the size of the cut and of the amount of material you will remove in a pass. You don't necessarily want to hog out a ⅜"-deep cut in a single pass, especially if you're using a router bit with a diameter of 1¼" to 1½".

When cutting half-laps on a table saw, you can use your regular combination blade or a dado stack. Guide the work using your miter gauge, a cutoff box, or a jig.

assembling lap joints

It's not difficult to assemble a frame joined with half-laps. You must apply clamps to the individual joints, however, in addition to using clamps that draw the assembly together.

Use bar or pipe clamps to pull the joints tight at the shoulders. Then squeeze the cheeks of individual joints tight using C-clamps or spring clamps.

►Gluing up a half-lapped frame requires the usual complement of pipe or bar clamps to pull the shoulders of the joints tight. Each joint also requires a C-clamp or spring clamp to pinch its cheeks together.

> Biscuit Joints

Biscuits can perform several functions. They can add strength to a joint, such as when you join a table apron to a leg. Or they can be used as an alignment aid, such as when you glue up a panel using several boards or you need to glue together veneered panels. The biscuits won't add strength here, but they will keep your parts in line as you clamp.

When making a biscuit joint, first put the two parts together and decide how many biscuits you need for that joint. A basic rule of thumb is to place your first biscuit 2" from the edge and then every 6" or so, though the spacing is really up to you. Draw a line across the joint at each spot where you want a biscuit. Set the fence on your biscuit joiner so the biscuit will be buried approximately in the middle of your material. (For example, if you're working with ¾"-thick wood, set your biscuit joiner for a ⅜"-deep cut. Don't worry about being dead-on in the middle. If you cut all your joints on one side, everything will line up.) Select the biscuit size you want to use—the biggest size biscuit that you can use is usually best—and dial that setting into your tool.

Clamp one of the parts to your bench. Align the line on the tool's fence or faceplate with the line on your work. Turn on the tool and allow it to get up to full speed. Plunge the tool into the wood and then out. Repeat this process for the other side of the joint.

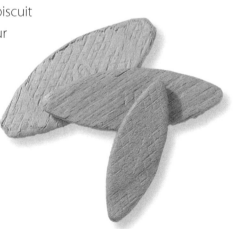

►The three most common sizes of biscuits: a No. 0 (the smallest), a No. 10, and a No. 20. Note: The biscuits pictured here are not actual size.

gluing up biscuit joints

There are at least two ways of gluing up biscuit joints. You can put glue in the slots and then insert the biscuit, or you can put glue on the biscuit and insert it in the slot.

For small projects, coat half the biscuit with glue and insert it into one of the slots. Then coat the other half of the biscuit and clamp your pieces together. This method produces clean joints with minimal glue squeeze-out, but it's a bit slow.

No matter which method you use, be sure to go easy on the clamping pressure. It's easy to distort a frame made with biscuits. If you're using a regular yellow glue, clamp the project for at least thirty to forty-five minutes before taking it out of the clamps.

where to use biscuits

Making the biscuit slot is easy. The tricky part is knowing when to use biscuits and how many biscuits to use. Here are some situations when you should be extra-careful:

▶ The two biscuits used here provide sufficient strength for attaching this apron and table leg.

Long-grain joints: Many people use biscuits to join several narrow pieces into a panel, such as a tabletop. Biscuits help align the boards so they don't slip as much when you clamp them. However, don't let anyone tell you that the biscuits make the joint stronger. In long-grain-to-long-grain joints, the glue is stronger than the wood itself. So biscuits here are only an alignment tool. Also, be careful to place the biscuits where they won't show after you trim your part to finished size.

Face frames: Biscuits are just right for face frames as long as your stock isn't too narrow. A No. 0 biscuit will work only with stock as narrow as 2⅜". Any narrower and the biscuit will poke out the sides. To join narrow stock, you need a biscuit joiner that can use a smaller cutter or a tool that cuts slots for mini-biscuits.

Continuous-stress joints: Biscuits are strong, but they are not a good choice for building a kitchen chair. The joints in chairs, especially where the seat meets the back, are subject to enormous amounts of stress. In such cases, a more old-fashioned choice like a mortise-and-tenon joint is a better idea.

With polyurethane adhesives: Polyurethane glue generally works quite well, but you must remember that biscuits swell and lock your joint in place by wicking up the water in white or yellow glue. Poly glues have no moisture in them. In fact, these glues need moisture to cure. If you want to use poly glue with biscuits, dip your biscuits in water before inserting them into the slot. The water swells the biscuit and activates the poly glue.

Building tables: If you're going to make a table using your biscuit joiner, use two stacked biscuits to attach the aprons and stretchers to the legs. This might mean making your aprons 7/8" thick. It's a good idea to add an extra biscuit whenever you're joining thick stock.

> Mortise-and-Tenon Joints

Cutting the mortise-and-tenon joint is one of the fundamental techniques you should master in order to graduate from "weekend woodworker" to "craftsman."

This joint is the foundation for projects that will stand up to use, abuse, and the march of time. It is the joint that ensures your chairs won't collapse under your weight. It reinforces doors so they'll remain square and tightly fitted. It resists racking so your workbench won't wobble.

To learn to make this joint, you first have to understand its parts and how to design the joint so it's strong and well-proportioned. At its most basic level, a tenon is merely a piece of wood that has been rabbeted on at least one—but usually all four—edges. The mortise is merely a hole that the tenon fits snugly into.

To design an effective joint, you need to make a tenon that is thick enough but not too thick, wide enough but not too wide, and long enough but not too long. And the same goes with the mortise. Here are some rules to follow:

Thickness: The rule of thumb is that tenons should be half the thickness of your work piece. This means a tenon on a piece of ¾" material should be ⅜" thick.

Width: You generally make the tenon as wide as possible so the wood resists twisting in its mortise. But once the joint gets close to 4" wide, you need to worry about wood movement. A joint wider than 4" will expand and contract enough to bust up your project. For that reason, when you are cutting tenons on wide stock, it is a good idea to make each tenon 3" wide and cut multiple tenons on the board. Leave a little extra space in the mortise to allow the tenons to move.

Length: This depends on your project. Typical casement tenons that are 1" long will be plenty strong. For large glass doors, make them 1¼" long. For small, lightweight frames and doors, stick with ¾"- or ⅝"-long tenons.

Edge shoulders: For most cases, edge shoulders (the distance from the edge of the tenon itself to the edge of the board it's on) should be ⅜" wide. Once they start getting larger than ½", you run the risk of allowing the work to twist.

Mortise dimensions: Designing the mortise is simpler. It should have the same dimensions as your tenon with one exception: Make the mortise ¹⁄₁₆" deeper than your tenon is long. This extra depth gives your excess glue a place to go and ensures your tenon won't bottom out.

methods of making mortise-and-tenon joints

There are an untold number of ways to make mortises and tenons. If you are using hand tools, you will be making tenons using a saw and chisel. Mortises can be made by hand using the thick chisels known (naturally) as mortising chisels.

If you are a power-tool woodworker, you are probably going to make tenons using your table saw, using a dado stack or a regular saw blade. One important thing to remember when cutting the shoulders of the tenon: Set the height of the blades to just a little shy of the shoulder cut you're after. You want to sneak up on the perfect setting by raising the arbor of the saw instead of lowering it.

For mortises, most woodworkers choose either a hollow-chisel mortising machine (similar to a drill press) made especially for making mortises, or a plunge router with a jig. If you use a router, remember to make multiple passes when routing your mortises, with no more than an ¼"-deep cut at a time.

►When making tenons with a dado stack in your table saw, the first pass should remove the bulk of the material. Keep firm downward pressure on your work, which will give you more accurate cuts.

►The second pass has the work against the fence and defines the face shoulder. There isn't any wood between the fence and blades, so the chance of kickback is minimal. The backing board reduces the chance of tear-out at the shoulders.

►Cut the edge shoulders and cheeks the same way you cut the face shoulders and cheeks.

> Dovetail Joints

One of the mysteries for beginning woodworkers is how to build dovetailed drawers. More than anything else, dovetails are seen as the mark of fine craftsmanship.

If you would like to make dovetails, a variety of jigs are sold that enable you to do so with a router and a bit that is shaped like a dovetail. You will need to follow the specific instructions that come with the jig. But, you may be asking, what about hand-cut dovetails?

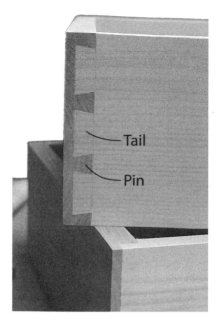

Tail

Pin

Making dovetails is not something simple that you can master with just a few tricks (sorry!). Cutting them is a skill that develops in time. The trick is to learn the basics and keep practicing.

In this section, you'll see the steps required to hand-cut *half-blind dovetails* (that is, dovetails that are only visible on one side of the corner). This is the type of dovetail most often used in drawer construction. For larger pieces, such as blanket chests, *through dovetails* are more common.

One excellent way to practice dovetails is to prepare two pine boards of equal widths. Cut your joints on the ends of each board and assemble them. Inspect the results and then trim off the joinery from each board and start dovetailing the boards again.

Save each joint you cut, though it's tempting to pitch bad examples in the trash. In fact, it's a good idea to put a date on each joint and stick it on a shelf in your shop. It's helpful to refer to your past mistakes so you can avoid future ones. Most of all, stick with it. Almost no one cuts good dovetails out of the gate. Ready? Let's go!

▶1) This is a source of debate among woodworkers, but in this example we are cutting the tails first. Set your marking gauge to ½" and scribe a line all the way around the front end of the side pieces.

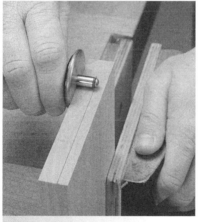

▶2) With your gauge at the same setting, scribe a line on the end grain from the back of the drawer front to locate how far the tails will penetrate into the drawer front.

▶3) Lay out the tails. Mark the center of the end grain on the sides. The tails shown here are 1¼" wide with 5/16" between each. Be sure to lay out your tails so the groove for the drawer bottom will be buried in a tail. Set your sliding bevel to 10°—a good angle for pine. Mark out the location of the tails on both faces with a knife.

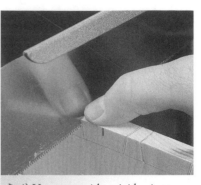

▶ 4) Use a saw with a rigid spine to cut the tails; cut to the outside of the line, not directly on the line.

▶ 5) Use a coping saw to remove the waste between your tails. First cut diagonally from corner to corner. Then come back and clean out the other piece of waste, as shown in the illustration.

▶ 6) Pare down the waste to your scribed lines with a ⅜" chisel. If your saw cuts are rough, use a ⅝" or ¾" chisel to shave the tails to your scribed lines. Do this in one quick cut if you can.

▶ 7) Now place your tails on top of the end grain of its mate. Using a sharp pencil or marking knife, scribe the outline of the tails onto the drawer front. This makes the outline for the pins.

▶ 8) On the inside of the drawer front, bring the mark for the end grain straight out to the scribed line. Cut these with a backsaw. You can make the cut 1" or so beyond your scribed line. This will allow you to clean out a lot more waste with your saw. Though not a routine method, it is a traditional one.

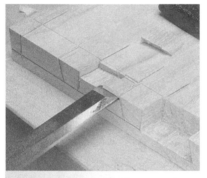

▶ 9) Remove the waste between the pins. First chop a line ¹⁄₁₆" in from your scribed line. Then bring your chisel back to your end grain and pop out the waste, as shown here.

▶ 10) Next, hold your chisel at an angle to pare out the waste down to your scribed line. Once you make these angled cuts, go back and hog out the waste with your chisel parallel to the scribed line but never crossing it.

▶ 11) Finally, fit the pins and the tails. To make fitting easier, use a knife to round over the edges of the tails buried in the drawer front. After fitting, use a dead-blow mallet to push the tails into the pins evenly. Congratulations!

Project 1: Page 105
Bathroom Shelf Unit

Project 2: Page 111
Handy Box

Project 3: Page 117
Desktop Organizer

Project 4: Page 123
Hanging Cupboard

Project 5: Page 129
Small Tool Chest

Project 6: Page 135
Sliding-Door Cabinet

Project 7: Page 141
Tall Bookcase with Drawers

Project 8: Page 147
Occasional Table

Hardwoods

RED ALDER

Alnus rubra

Color: from off-white to pale-pink brownish color

Grows: on the Pacific Coast from Alaska to California

Properties: moderately lightweight; of medium strength; low shock resistance

Uses: mostly furniture, also for millwork

BALSA WOOD

Ochroma lagopus

Color: pale beige to pink

Grows: South and Central America

Properties: the lightest commercial hardwood

Uses: insulation, model-making, packaging

WHITE ASH

Fraxinus americana

Color: heartwood is brown; sapwood is off-white

Grows: in eastern United States

Properties: heavy; stiff; strong; shock resistant

Uses: mostly for oars, sporting implements such as baseball bats, decorative veneer, furniture, and cabinetry

BASSWOOD

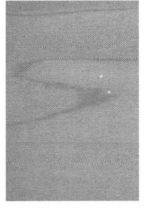

Tilia americana

Color: heartwood is yellow-brown with dark streaks; sapwood is white or brown

Grows: eastern United States, from Canada south

Properties: soft; lightweight; straight-grained and easy to work

Uses: molding and woodenware, including woodcarving and veneers

ASPEN

Populus grandidentata

Color: heartwood is off-gray/brown; sapwood is lighter in color

Grows: in the Northeast

Properties: uniform texture; easy to work; lightweight

Uses: mostly furniture, also for millwork, particleboard, and veneer

BEECH

Fagus grandifolia

Color: heartwood is reddish brown; sapwood is white

Grows: eastern United States

Properties: heavy; hard and strong; good for steam-bending and woodturning

Uses: for flooring, veneer, containers, and furniture making

Hardwoods

YELLOW BIRCH

Betula alleghaniensis

Color: heartwood is reddish brown; sapwood is white

Grows: in the Northeast and Great Lakes States, and Appalachian Mountains

Properties: heavy, hard, and strong with uniform texture

Uses: furniture, boxes, doors and other interior woodwork, and veneer

ELM

Ulmus americana

Color: heartwood is light brown with red; sapwood is off-white

Grows: in eastern United States

Properties: moderately hard, stiff and heavy; good for bending

Uses: mostly veneer for furniture and decorative panels

BUTTERNUT

Juglans cinerea

Color: heartwood is light brown; sapwood is off-white

Grows: from Canada to Maine, west to Minnesota

Properties: lightweight; coarse in texture; machines and finishes easily

Uses: furniture, cabinets, veneer, and interior woodwork

CHESTNUT

Castanea dentata

Color: heartwood is gray-brown; sapwood is off-white

Grows: previously in eastern United States; 1920s blight killed almost all; today, comes mostly from salvaged timbers

Properties: lightweight, little strength, easy to work

Uses: furniture, caskets and veneer core stock; mostly interior woodwork

CHERRY

Prunus serotina

Color: heartwood is reddish brown; sapwood is white

Grows: southeastern Canada, eastern United States

Properties: strong and stiff; uniform in texture; machines well

Uses: furniture, veneer, caskets, architectural woodwork

HICKORY

Carya ovata

Color: heartwood is red; sapwood is white

Grows: eastern, central, and southern United States

Properties: heavy; hard and strong

Uses: mostly for tool handles, dowels, and some furniture; also for flavoring meat

Hardwoods

SOFT MAPLE

Acer saccharinum

Color: heartwood is light reddish brown (lighter than hard maple); sapwood is white with reddish tinge

Grows: in eastern United States

Properties: similar to hard maple, but not as heavy, hard, or strong

Uses: veneer, some furniture, railroad crossties

YELLOW POPLAR

Liriodendron tulipifera

Color: heartwood is yellow-brown; sapwood is white

Grows: in east, south, and midwestern United States

Properties: straight-grained; uniform in texture

Uses: mostly for furniture, molding, cabinets, and musical instruments; also used for plywood

HARD MAPLE

Acer saccharum

Color: heartwood is light reddish brown; sapwood is white with reddish tinge

Grows: in eastern United States and Great Lakes States

Properties: heavy; strong, stiff, and shock resistant; grain is straight, but bird's-eye, curly, and fiddleback maple add decorative qualities

Uses: lumber, veneer, furniture, cabinets, cutting boards, bowling alleys, and bobbins

WHITE OAK

Quercus alba

Color: heartwood is gray-brown; sapwood is white

Grows: in the South Atlantic and central states

Properties: heavy and very impermeable to liquids

Uses: cooperage, veneer, shipbuilding, furniture, and millwork

RED OAK

Quercus rubra

Color: heartwood is reddish brown; sapwood is off-white

Grows: in eastern United States

Properties: heavy; quartersawn lumber evident by broad rays in grain

Uses: lumber, veneer, millwork, furniture, and caskets

SASSAFRAS

Sassafras albidum

Color: heartwood is dull brown; sapwood is light yellow

Grows: in eastern United States, and from southeastern Iowa to eastern Texas

Properties: moderately hard and heavy; resistant to decay

Uses: small boats, fence posts, and millwork

Hardwoods

PECAN

Carya illinoensis

Color: heartwood is reddish brown with dark brown stripes; sapwood is white or creamy white with pinkish tones

Grows: southern United States

Properties: open grain; occasionally wavy or irregular

Uses: chairs and bentwood furniture

BLACK WALNUT

Juglans nigra

Color: heartwood is light to dark brown; sapwood is off-white

Grows: from Vermont south and to the Great Plains

Properties: straight-grained and easy to work; heavy; hard and stiff; good for natural finishing

Uses: valued for furniture and architectural woodwork because of its interesting grain patterns; also used for gunstocks and interior woodwork

SWEETGUM

Liquidambar styraciflua

Color: heartwood is reddish brown; sapwood is light colored

Grows: from Connecticut to Missouri and south to the Gulf Coast

Properties: moderately heavy and hard

Uses: quartersawn pieces are great for furniture and woodwork, also used for veneer and plywood

TUPELO

Nyssa aquatica

Color: heartwood is light brown; sapwood is lighter colored

Grows: in the southeastern United States

Properties: interlocked grain; uniform in texture; moderately heavy and strong

Uses: lumber, veneer, furniture, pallets, and crates

SYCAMORE

Platanus occidentalis

Color: heartwood is reddish brown; sapwood is lighter colored

Grows: from Maine to Nebraska, south to Florida and Texas

Properties: interlocked grain and nice texture; moderately hard and heavy

Uses: lumber, veneer, fuel, and furniture; also small boxes and butcher blocks

WILLOW

Salix alba

Color: heartwood is gray-brown; sapwood is creamy yellow

Grows: in the Mississippi Valley

Properties: uniform in texture; slightly interlocked grain; lightweight

Uses: lumber and veneer, as well as furniture, boxes, and caskets

Softwoods

EASTERN RED CEDAR

Juniperus virginiana

Color: heartwood is red; sapwood is off-white

Grows: in eastern United States

Properties: moderately heavy and strong; uniform in texture; often has many small knots; resistant to decay

Uses: mostly fence posts, but also chests, wardrobes, and small wood projects like pencils

CYPRESS

Taxodium distichum

Color: heartwood varies from yellow to dark brown; sapwood is mostly white

Grows: in the southern United States

Properties: moderately hard, heavy, and strong

Uses: siding, millwork, and interior woodwork

WESTERN RED CEDAR

Thuja plicata

Color: heartwood is reddish brown; sapwood is off-white

Grows: from the Pacific Coast to Alaska

Properties: straight-grained; uniform in texture; lightweight; resistant to decay

Uses: interior woodwork; shipbuilding; doors

DOUGLAS FIR

Pseudotsuga menziesii

Color: heartwood is reddish or yellow; sapwood is lighter in color

Grows: from Mexico to the Pacific Coast

Properties: varies widely

Uses: lumber, plywood, general millwork, and sometimes furniture

WHITE CEDAR

Thuja occidentalis

Color: heartwood is light brown; sapwood is mostly white

Grows: from Maine south along the Appalachians and west to the Great Lakes States

Properties: lightweight; low in strength; works easily

Uses: cabin logs, woodenware, boats, fencing

HEMLOCK

Tsuga heterophylla

Color: heartwood and sapwood are both white, with a reddish-purple coloring

Grows: along Pacific Coast and in Rocky Mountains

Properties: lightweight and moderately strong and hard

Uses: plywood and lumber for furniture, boxes, and ladders

Softwoods

PONDEROSA PINE

Pinus ponderosa

Color: heartwood is light red and brown; sapwood is mostly white

Grows: in the western United States

Properties: moderately lightweight and soft; uniform in texture; straight-grained

Uses: lumber and veneer; some interior woodwork

SITKA SPRUCE

Picea sitchensis

Color: heartwood is light brown; sapwood is white

Grows: along northwest coast from California to Alaska

Properties: moderately lightweight and soft; uniform in texture

Uses: mostly lumber, pulpwood, furniture, millwork, and aircraft construction

REDWOOD

Sequoia sempervirens

Color: heartwood is light red to deep brown; sapwood is mostly white

Grows: on the coast of California

Properties: straight-grained; easy to work; fairly decay resistant

Uses: mostly construction and outdoor furniture

WHITE PINE

Pinus strobus

Color: heartwood is off-white to reddish; sapwood is off-white

Grows: mostly in Idaho and Washington

Properties: straight-grained and easy to work

Uses: lumber, millwork, interior woodwork

SUGAR PINE

Pinus lambertiana

Color: heartwood is light brown; sapwood is white

Grows: in California and Oregon

Properties: uniform in texture; straight-grained; easy to work; good for nailing

Uses: general millwork, boxes, doors, and frames

SOUTHERN PINE

four major species

(Pinus palustris, longleaf; Pinus echinata, shortleaf; Pinus taeda, loblolly; and Pinus elliottii, slash pine)

Color: heartwood is reddish brown; sapwood is off-white

Grows: throughout eastern and southern United States

Properties: longleaf and slash are heavy, strong, and hard; shortleaf and loblolly are more lightweight

Uses: structural-grade plywood, extensive range of construction uses, interior woodwork, boxes, and pallets

Imports

BUBINGA

Guibourtia

Color: ranges from light red to brown with purple streaks

Grows: in West Africa

Properties: moderately hard and heavy, but fairly easy to work

Uses: furniture, veneers, and turning

PURPLEHEART

Peltogyne

Color: brown when first cut, then turns a deep purple, then a dark brown over time

Grows: from Mexico to Brazil

Properties: moderately difficult to work; good for turning; takes finishes well

Uses: turning, marquetry, fine furniture, some shipbuilding

PADAUK

Pterocarpus soyauxii

Color: deep red-brown

Grows: in Africa

Properties: coarse in texture with interlocking grain; easy to work; finishes well

Uses: veneer, fine cabinetry, and other woodwork

HONDURAN MAHOGANY

Swietenia macrophylla

Color: light pink to dark reddish brown

Grows: Mexico, Central America, and South America

Properties: fine to coarse texture; decay resistant; very easy to work; takes finish well

Uses: fine furniture, musical instruments, veneers, turning, and carving

JELUTONG

Dyera costulata

Color: whitish

Grows: in Malaysia

Properties: straight-grained; moderately uniform in texture

Uses: making patterns, wooden shoes, and picture frames

BRAZILIAN ROSEWOOD

Dalbergia nigra

Color: heartwood is brown or purple with black streaks; sapwood is white

Grows: in Brazil

Properties: coarse; straight-grained; oily; hard; heavy and strong

Uses: mostly plywood veneer, turning

Imports

LUAN MAHOGANY

Shorea

Color: extremely varied

Grows: in Southeast Asia

Properties: part of the meranti group of wood; interlocked grain

Uses: furniture, cabinetry, molding, and some decking

WENGE

Milletia laurentii

Color: deep brown with black striping

Grows: in Africa

Properties: hard; coarse in texture; finishes well

Uses: fine cabinetry, fine furniture, woodturning

INDIAN ROSEWOOD

Dalbergia latifolia

Color: heartwood is brown to purple with dark streaks; sapwood is yellowish

Grows: in India

Properties: heavy and strong; moderately difficult to work; good for turning

Uses: mostly veneer

ZEBRAWOOD

Microberlinia brazzavillensis

Color: light brown with dark striping

Grows: in Africa

Properties: hard; heavy and coarse in texture with interlocking grain; finishes well

Uses: as inlay for decorative color contrast

TEAK

Tectona grandis

Color: yellow to dark brown

Grows: India, Myanmar, Thailand, Laos, Cambodia, Vietnam, and the East Indies

Properties: coarse texture; straight-grained; oily; moderately easy to work

Uses: shipbuilding, furniture, and decorative veneer

chapter 7

Setting Up Your Shop

Now that you know about all these great tools, materials, and techniques, you may be anxious to get out to the garage and start working on that end table.

First, though, you should learn how to set up your shop the right way. Setting up your workspace properly is an essential step toward a productive and enjoyable woodworking experience. Whether it's a garage, basement, large shed, or open space, and whether you're working with hand tools, benchtop machines, or floor models, properly placing and storing all of your tools and machines is ultra important.

As you get more involved with woodworking, the number of tools that you simply can't live without will multiply. And the number of leftover scraps of wood that you "just might need someday" will pile up even faster. Having the right spaces for tool and material storage, and keeping your shop clean and organized, will then become even more important if you want your workshop to be a safe and productive space.

> Planning Your Shop Space

A good way to organize where you're going to place everything in your shop is by plotting it out on paper. This will be easier on your back than lifting heavy equipment and lugging it around! Draw your shop on graph paper. Then arrange the tools in ways that work best for you. Which tools do you use most? Which do you use least? In what area of your shop do you work the most and what do you need to have handy while you are working in that space? Remember what you do with each machine. For example, with a table saw, you'll need to maneuver a 4' × 8' sheet of plywood to the back, front, and left side of the saw. The following diagram shows you the necessary working footprint for each major machine. Photocopy and cut out the tools from the diagram (those that you own now or plan to own eventually, that is) to start seeing what combination works best for your workspace.

Don't worry if many—or even all—of the large tools and machines shown here seem beyond your reach, at least for now. Woodworkers with limited space—or a limited budget—won't be able to own many big stationary tools. Hand tools and small power tools will make up your woodworking arsenal. If you can afford to have one stationary tool, though, your best bet is a table saw; as we've seen, table saws can be used in so many woodworking applications that they are truly the center of any woodworking shop, big or small.

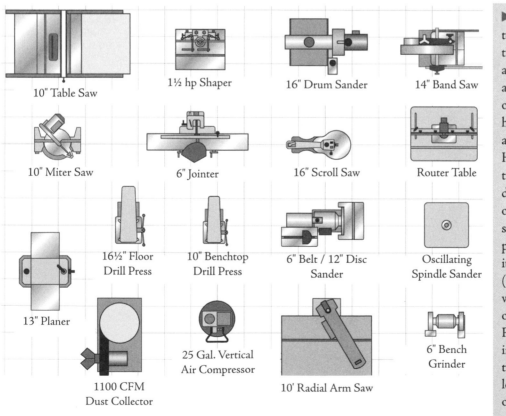

10" Table Saw

1½ hp Shaper

16" Drum Sander

14" Band Saw

10" Miter Saw

6" Jointer

16" Scroll Saw

Router Table

16½" Floor Drill Press

10" Benchtop Drill Press

6" Belt / 12" Disc Sander

Oscillating Spindle Sander

13" Planer

1100 CFM Dust Collector

25 Gal. Vertical Air Compressor

10' Radial Arm Saw

6" Bench Grinder

▶You can copy this template, cut out the relevant pieces, and move them around on a piece of graph paper that has ¼" squares as are shown here. Here, each square translates to 1', so draw the outlines of your working space on the graph paper according to its actual dimensions (that is, an 8' wall would be drawn over eight squares). Remember to include space around the machines for the length of the wood on the in-feed and out-feed sides.

> Setting Up the Workflow

What you should also know about setting up shop is that you need to create an orderly flow of work from raw lumber to the finished product. The workflow always starts from where the wood is stored, or where it enters the workshop. Next, the lumber is prepared for use by jointing, planing, and sawing to the proper dimensions. Conveniently, the machines required for these steps are also the ones that need the most power and create the most dust, allowing you to locate your power and dust collection in a "machining" area, with these machines close to one another.

From the machining phase, move to joinery and assembly, usually requiring hand tools, a band saw, drill press, and hand-held power tools, such as a router, biscuit joiner, and brad nailer. A stable workbench or assembly table is ideal.

The assembly area should be located out of the way of the machining area (if you have enough space). Your hand and small power tools should be stored in easily accessible drawers or on the wall, and quick access to clamps will make things easier too.

Once the assembly is complete, the third phase is finishing. No matter what finish you use, a clean, well-ventilated area is required.

When applying a varnish or shellac finish, the vapors that are created are flammable and should be kept away from any ignition points, such as water or space heaters. In concentrated exposure, the vapors can also be harmful to you, so ventilation is important. In addition, when storing solvent-based finishes (like varnishes), a fireproof storage cabinet is a must.

If you're going to use a spray-on finishing system, ventilation is even more critical to move the over-spray away from your lungs.

From here, the rest of your shop will fall into place. To save on space, store your tools under cabinets until you need them.

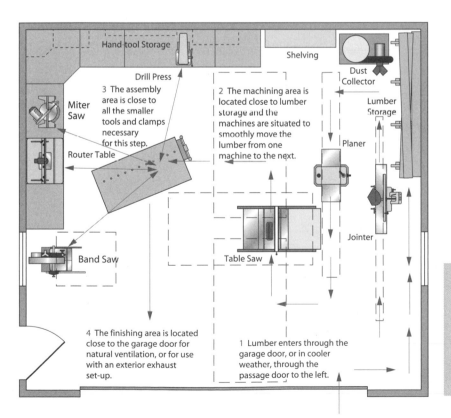

Hand-tool Storage

Shelving

Dust Collector

Drill Press

Miter Saw

3 The assembly area is close to all the smaller tools and clamps necessary for this step.

2 The machining area is located close to lumber storage and the machines are situated to smoothly move the lumber from one machine to the next.

Lumber Storage

Router Table

Planer

Band Saw

Table Saw

Jointer

4 The finishing area is located close to the garage door for natural ventilation, or for use with an exterior exhaust set-up.

1 Lumber enters through the garage door, or in cooler weather, through the passage door to the left.

► This is an example of workflow in a two-car garage turned workshop. Follow the red arrows to the easiest path for woodworking. The dotted lines show approximate in-feed and out-feed room for lumber.

walls, floors, and lights

As we mentioned in the chapter on shop safety, most woodworkers don't think about painting the walls of their shop, but it is obviously a good idea to do so before you start covering them with cabinets or shelves, and before you fill up the floor with heavy machinery. A coat or two of paint will help keep out moisture and make it easier to wipe off accumulations of dust. When choosing a paint color, remember that lighter colors reflect light better—a helpful quality in spaces that often have few windows, and where you really need to be able to see what you're doing.

Your workshop floor will need to be level, dry, and easy to sweep and clean. You should also realize that woodworking involves long periods of standing on your feet. As we also mentioned in Chapter 5, putting down rubber mats in the areas where you'll be standing a lot can make things easier on your feet and legs, as well as provide better traction for safety.

You'll also need to keep in mind your lighting needs and your power outlets and what they can handle. Some specific tasks in the woodshop (and whether you're right-handed or left-handed) might require additional or adjustable lighting. You need to see what you're doing! You might want to think about setting up lamps as side lights, or using portable light stands that you can move around depending on the project you have going and its stage of production.

> Storing Materials and Tools

When choosing storage options, you'll have to decide whether you need drawer cabinets, door cabinets, or both. If you're storing large, odd-shaped items (like a belt sander or arc welder), a drawer can be a real problem. A door cabinet is a better place to store bulky items.

If you're storing smaller items (like door hinges, glue, or seldom-used jigs) a door cabinet can be a great place to lose these items. Beyond doors or drawers, you have two general choices in cabinets: buy them or make them. If you make your own cabinetry, you will almost certainly get exactly what you need for the best space utilization. You'll also most likely save money, but constructing them will take up a chunk of your time.

Think about purchasing shop-grade cabinets from a home improvement center. There are a number of utility cabinet models available in various shapes and sizes. Aside from cabinets, shelves are a good storing option. Keep in mind that you can easily find what you're looking for with shelves, but that everyone else can see all that you have placed there—attractive or not. Be sure you know what weight your shelves can support to prevent collapses.

▶The cabinets for your shop can be premade kitchen cabinets, cabinets you make yourself, or cabinets designed for your woodworking needs.

wood storage

There are three types of wood stored in a shop: sheet goods (plywood), rough or full-size lumber, and shorts and scraps. Shorts and scraps are the pieces you can't bring yourself to throw away. Not only are there usually more of these pieces, but they're also harder to store than plywood or rough lumber because of their odd shapes and sizes.

Plywood takes up the least amount of space when stored standing on edge. Most of us aren't storing more than a few sheets of plywood, so this can often be stored in a 10"-to-12"-deep rack that can slip behind other storage or machinery.

Rough lumber is best stored flat and supported to keep the wood from warping. Keeping it up off the floor also keeps it away from any water that may get into your shop. A wall rack with a number of adjustable height supports provides the easiest access while keeping the wood flat and dry.

Shorts are the most difficult to store, but a rolling box with a number of smaller compartments holding the shorts upright allows easy access to the pieces, and it keeps them from falling against and on top of one another.

tool and material storage

There is nothing more frustrating than not finding a tool you need at a time when you really need it. For this reason, hammers and other hand tools are usually best hung on the wall. There are many ways you can hang these tools. The most common material used to hang tools is the pegboard. It's inexpensive, versatile, and easy to mount.

You can also use slat wall. This type of hanging system is used for displaying socks and ties at department stores. Slat wall costs more than pegboard.

Feel free to get creative and use other materials like drywall molly bolts to hang tools. Trace the outline of the tool on the board or wall that you create in order to ensure that you put the tool back in its proper place.

Finishing materials deserve special attention. While waterborne finishes are gaining in popularity, flammable finishes in cans, bottles, and jars should be stored in a fireproof storage box, which should be kept clean and organized at all times. A tall cabinet with adjustable shelf space is ideal.

Other workshop storage needs fall into the cabinet and shelving category. Just because there's a tool sitting on the floor against the wall does not mean you can't hang a cabinet or shelving above it. In fact, in many cases there are accessories and supplies you need near that tool that belong on a shelf right above it. And don't hesitate to go all the way to the ceiling with storage. Even though the top shelves are harder to get to, we all have things that aren't often used.

▶This shop has well-defined areas for wood and tool storage—and a number of clamps ready for immediate use.

Planning Your Projects

Your shop is all ready to go. But what are you going to make in it? That's what this chapter, and the eight projects in the rest of the book, are all about.

First, we'll take a very quick tour through the history of furniture design. Recognizing the names and common features of some of the most prominent furniture styles will help you to develop your own preferences for what it is you want to create. A Chippendale chair may be a little out of your reach right now, but a Shaker side table might be just the thing.

Then we'll discuss some of the things you need to know to make patterns and to design furniture. Like many woodworkers, you may begin by making projects that closely follow the pattern in a magazine or a book such as this one. Eventually, though, you will probably want to adapt projects to fit your particular needs—the size of your dining room, the wood you have available, the joinery techniques you prefer, or your favorite furniture style. After that, you may even want to make projects entirely of your own design. In that case, you'll discover the satisfaction that comes from knowing that you've not only built something, but also imagined it.

> A History of Furniture Styles

If you're really ambitious, or just plain curious, you can track furniture styles back to the ornate Byzantine-era religious furniture or even back to the clunky furniture of the Middle Ages. But most modern woodworkers stick to more recent influences, from about the 1700s on. Here are some brief descriptions of furniture styles you may choose to merely admire from afar, or employ as inspiration for your own woodworking designs.

eighteenth-century styles

Queen Anne of England was in charge from 1702 until 1714, although the *Queen Anne* style of furniture was not influenced by her and actually incorporates furniture designed through the mid-1700s.

One telltale element of a Queen Anne furnishing is the use of the cabriole leg, which is utilitarian in its solid strength (although sometimes made thin and delicate), but also made attractive by its curves and pad feet.

You'll often see rosette and leaf carvings on Queen Anne furniture, as well as butterfly drawer pulls, finials, and frontal veneers. The Queen Anne style, in general, is marked by graceful lines and curves, and it often features furniture made of walnut, cherry, or maple woods.

Thomas Chippendale, who lived from 1718 to 1779, was an English cabinetmaker and woodcarver whose furniture designs blended the Queen Anne style with more radical elements, such as Chinese influences. His style, naturally known as *Chippendale* style, became very popular with the wealthy and with Americans after he published a trade catalog, *Gentleman and Cabinet Maker's Director,* in 1754.

Queen Anne elements of Chippendale furniture include use of the cabriole leg, as well as ball-and-claw feet and fancier French rococo carving elements. Chinese Chippendale furniture was very different from the other furniture he designed, and somewhat modern in appearance, with geometric frameworks in chair backings.

Another English furniture designer, Thomas Sheraton Jr., who lived from 1751 until 1806, designed pieces similar to his designing predecessor George Hepplewhite. The *Hepplewhite* and *Sheraton* styles are very similar, and borrowed heavily from Greek and Roman classical motifs.

Sheraton's style differed with his creative blending of utilitarian and aesthetically pleasing elements, in pieces like his twin beds, small tables, and shaving mirrors.

British and American furniture styles weren't very different until the obvious division that occurred as a result of the American Revolution. American cabinetmakers and furniture designers began creating their own styles at the end of the eighteenth century, which

▶A cabriole leg from a chest made in the Queen Anne style

►Detail of a
Chippendale
cabinet

►A bed made in
the Sheraton style

became known as the *Federal* style. Popular furniture of that time, from about 1790 until about 1830, actually was related to Hepplewhite and Sheraton furniture and displayed many neoclassical elements, with a French-American twist.

arts and crafts style

Falling at the tail end of the Victorian Age's historic revival furniture, which enjoyed popularity from about 1830 until shortly after the turn of the century, *Arts and Crafts* furniture gained momentum partially in opposition to the opulent and somewhat frivolous Victorian furniture.

Arts and Crafts furniture, sometimes categorized as *Craftsman* or *Mission* furniture, was a movement and a style focusing on the importance of simple, handmade furniture, spearheaded by William Morris in England and Gustav Stickley in the United States. Often made of oak, Arts and Crafts furniture includes square designs, exposed mortise-and-tenon joinery, slat backs, and, sometimes, leather upholstery.

shaker style

Shakers relocated from England to the United States in the late 1700s and established early settlements in New York, New England, and the Midwest. Austere and utilitarian in nature, due to the strict religious beliefs of the Shaker people, *Shaker* furniture is now ironically very sought after and admired in general society. Its beauty lies in its simplicity.

►A rocking chair in the Arts and Crafts style

Simple lines, elegant turnings, plain feet, and oval boxes are characteristic of Shaker furnishings. Made mostly of pine and maple, Shaker furniture also consists of cane-seat ladder-back chairs, simple candle wall sconces, tall cupboards, or chests with wooden knobs, as well as clocks, tables, dressers, and desks.

It seems to hold true even for non-Shaker fans of this style that, according to Shaker belief, "That which has the highest use possesses the greatest beauty."

modern styles

"Modern" is a slightly ambiguous term when it comes to classifying furniture styles. It basically encompasses all styles that have existed from about 1900 to the present, which is a wide range of styles difficult to categorize.

▶A cabinet in the Shaker style

The *Bauhaus School* style (which was popular from about 1919 to 1933) incorporated minimalism and functionalism, and has had a serious effect on modern architecture and design. *Art Deco* furniture became popular in the 1920s and 1930s, as did an entire Art Deco style of architecture and design. Characterized by streamlined, geometric forms, Art Deco furniture includes rounded fronts and wood mixed with chrome hardware, glass, or other materials.

> Making Patterns

When designing your own furniture, the style and method options are endless. You may know that you like Arts and Crafts furniture, for example, and decide to implement certain characteristics from that style into your own designs. Or maybe you just see something you like in a magazine and want to try to make your own at home, rather than buying it from a store.

Whichever way you go, you're going to have to sketch out what you want before you do anything else. You will usually also want to make a pattern to follow.

A pattern is the full-size outline of the object that you are going to make, drawn on paper or a template board. Even furniture with unique curves and fancy outlines can be accomplished with a little bit of technique.

For patterns that you want to transfer from magazines or other photographs, you can do this through the use of large squares drawn on a sheet of paper. Draw the corresponding lines through the large squares in the same way they go through the small squares (on the original). After this is done, the pattern can be transferred to the template or master pattern by tracing it on carbon paper over the template. Be sure your pattern is secure to eliminate shifting or slippage.

Most patterns can be cut out with a band saw, jigsaw, or coping saw. Band saws seem to work best, if you have one, because they make it easy to cut out the project's more minute details.

The Sheraton style of furniture provides an example of a rounded-leg curved pattern you might come across or use. The legs can be stacked together and cut in multiples of two or four, depending on your saw and its size. The final template should be sanded smooth for quality in your duplications. This would be your final pattern for that part of the project, so it should be as exact as possible.

One thing to note when cutting your pattern: Keep the blade on the outside of the line of the pattern, since you can always sand it or file it down, or saw some off, if you have to—but you can't add on to it!

It's also a good idea to use a ¼" underlay with your template. It can be readily cut on a band saw (or jigsaw) and can be sanded quickly or recut for final dimensioning. The templates are easy to store for later use if the need arises.

> Standard Guidelines for Furniture

Designing furniture involves much more than feverishly sketching an artistic form onto a scrap of paper, holding it up, and shouting victoriously, "Yes! That's it! The most beautiful table anyone will ever build!"

No matter how inspired your design is, it won't lead to a functional piece of furniture unless you follow the most basic dimensions and standards that other woodworkers, through trial and error, have discovered make the most sense and result in the most sturdy, useful final projects. The following sections will give you many of the dimensions and features to consider when making furniture.

guidelines for shelving size and construction

Shelves can vary in size from a 1' wall mount to 60" to 78" of floor shelves.

Keep in mind the reasons you have for building a piece, as well as the aesthetics you desire. If you're building a bookcase, for example, particleboard might not be the best material in terms of sag and durability. Although particleboard is laminated on both sides and is pretty thick, the weight can add up quickly. Hardwood seems to be the best choice for avoiding sag when building shelves.

Bookshelves are fairly uniform in size (for example, often 12" high between shelves and 11½" deep), but you can vary size to accommodate individual needs. If your length is going to exceed 36", it's a good idea to use a horizontal cleat for added strength. The back can be made out of lighter material, but you can also rabbet your backboards for a square, flush fit. If you plan to use shelves for heavier books or objects, use a heavier plywood for the back and rabbet the backboard to keep the shelves snug and to avoid the possibility of sag.

Whether you're building shelves that will be used to hold CDs, magazines, photo albums, or clothes, first measure the depth, height, and width you plan to place the shelf unit in, thinking all the while about the décor of the rest of the room or unit.

What you're building becomes a piece of furniture once it's sitting in view, so take a few minutes to think about how you plan to give it some visual appeal, and not just the practical aspects.

If the sides and shelves are made of a high-quality wood, then you'll want to make the face from a top-quality wood, also. This will conceal any exposed edges to give your final project a finished look. Cleats attached to the front can also be appealing to the eye, if done in a smooth manner with detailed design. Feel free to use screws for greater strength anywhere they can be installed and not exposed to view.

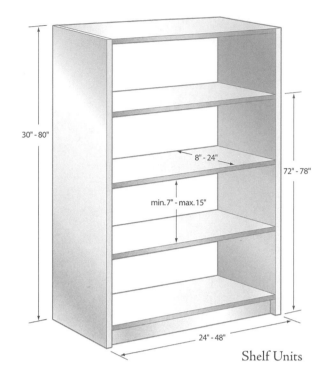

Shelf Units

Some possible joint choices for shelf construction:

◆ Butt joint: easy joint; use glue and nails

◆ Cleat: simple joint; unsightly but economical

◆ Full dado: simple joint; fairly strong, but needs facing if you don't want the dado to be visible

◆ Blind dado: fairly intricate joint; very strong

◆ Sliding dovetail: strong and exact joint, but not easy to make

◆ Biscuit: simple joint (if you own a biscuit joiner tool); good for general shelf use

Sizes for Common Shelf Storage Objects

	Object	Depth × Height (inches)
Books	Paperback	4¼ × 6⅞
	Standard hardback	7 × 9½
	Large book (e.g., textbook)	9 × 11
	Art/coffee table book	11 × 15
Music	Vinyl LP	12⅜ × 12⅜
	Compact disk	5½ × 5
	Audiocassette	2¾ × 4¼
Video	DVD	5½ × 7½
	VHS videocassette	4⅛ × 7½
	8mm VHS	4⅛ × 7½
Computer	CD-ROM	5½ × 5
	Floppy disk	3½ × 3½

When building bookcases, entertainment units, or any piece of furniture that is meant to hold things, use these handy dimensions (above) to figure out the needed space for these common items.

guidelines for tables and chairs

Most tables, whether dining or card, are constructed to accommodate the average-size person. The most common table height is 29", but, as the sizes of humans vary, as well as the purposes of tables, that 29" number varies, as well.

Tables basically vary in height from 26" to 32". Drafting tables and kitchen bar tables are higher, which means the stools designed to accompany them are higher, as well.

If you are constructing a new card table or a new dining table for a particular home or family, take into consideration the general height of its members, and make the table and chairs taller or shorter for comfort.

The general rule is to allow 12" space for a person's knees and 6" to 9" for leg clearance for each person sitting at the table. Try to allow at least 24" to 26" of space along the edge of the table for each individual. Think about the last wedding reception you went to: They always seem to crowd a lot of uncomfortable guests at rectangular tables, but people sitting at crowded round tables don't seem as uncomfortable. Round tables seem to provide a little more room for each person.

When thinking about the chairs that will accompany a table, be aware that the taller the chair back, the more formal the chair appears. Also, more room is needed for formal tables, as chairs are often wider and may have armrests. For those situations, plan for at least 2" extra on each side.

The necessary circumference of a round table can be calculated by multiplying the number of persons seated by 24" or 26", depending on style and chairs. Calculate the diameter by dividing the circumference by pi (3.14), and there you have it.

One more thought about table design: putting in a little extra room, if you can, will make those Thanksgiving dinners when all the extra chairs are pulled up to the table a little more enjoyable.

Here's a listing of typical table sizes—all in this one handy table!

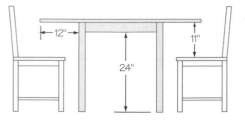

Average Table Sizes (in inches)

Table Type	Length	Height	Width
Square card	32–40	28–30	32–40
Round card	48–50 dia.	28–32	
Dining	60–96	28–30	40–42
Oval dining	42–56	28–30	40–42
Workbench	36–?	32–36	24–28
Sofa table	48–72	24–30	16–20
Coffee table	⅔ length of sofa	16–18	36–56
End	22–30	16–24	16–24
Nightstand	16–24	16–26	16–24

a last note on furniture construction and project size

One thing to keep in mind when constructing furniture of any sort in your workshop or basement (especially your basement) is size.

Don't learn the hard way that you have to be able to get your final project up the stairs and through the shop doorway.

Most woodworkers know at least one other woodworker who has done this: spent months perfecting a hutch or cabinet or dresser, only to realize upon completion that it won't fit through the shop door.

Try to always put a 36" door in any workshop, or even larger if space allows. If you have the luxury or the forethought, create a workshop space ideal for cutting long boards and allowing for the mobility of large objects. A 32' × 16' shop would be ideal for ample maneuvering space, although any shop can work well as long as you use your space efficiently.

Part Two

Woodworking Projects

Project 1:	Bathroom Shelf Unit
Project 2:	Handy Box
Project 3:	Desktop Organizer
Project 4:	Hanging Cupboard
Project 5:	Small Tool Chest
Project 6:	Sliding-Door Cabinet
Project 7:	Tall Bookcase with Drawers
Project 8:	Occasional Table

The following eight projects are roughly arranged in order from easier to more difficult. How difficult a particular project will be for you depends on several factors, such as what tools you own and what woodworking techniques you've mastered. A few things to know before you pick one of the projects and dive in:

Materials Lists

The "Materials List" that accompanies each project will give the quantity and dimensions for all the pieces of wood that you will need to cut. (This part of the list is often called a "cutting list.") When buying wood for a project, it is best to add up the amount of wood in the cutting list, and then add on 25 percent or so to the total. You should then have enough to make up for wood lost in trimming to size, or in cutting to avoid knots and other imperfections.

The end of the Materials List gives *Other Materials* you will need—metal fasteners, biscuits, hardware, etc.—that don't need to be cut. This list won't include some of the materials that you will need to complete the project, such as glue and sandpaper. Those are things that every shop should always have on hand.

Finishes

Some of the projects give detailed information on what finish to apply; others give more general instruction. The exact finish you use is always up to you; take into consideration the type of wood you are finishing, and what the piece will be used for.

Wood Choices

The Materials Lists will tell you the types of woods that were used to make each project as it is pictured. You may decide to use a different wood, but first make sure that the new wood you choose is right for the piece you're making. For example, the Tall Bookcase with Drawers on page 140 is made of white oak, a wood commonly used in the Craftsman-style furniture that is the inspiration for the piece. You can make it with a different wood, but your finished piece won't carry the same historical connections that one in oak would.

Joinery Choices

When you are considering making one of these projects, you should first look at what types of joinery it includes and what tools you will need. You will not get far in making a project full of biscuit joints if you don't own a biscuit cutter.

As you get more confident in your woodworking, you can make more of the joinery decisions yourself and adapt instructions to your preferences and skills. You should make sure that you take everything into consideration when doing so, though. If you are making a bookcase and you switch from a biscuit joint to a dado joint, for example, you will need to increase the length of the shelves by the total depths of the two dado grooves.

Safety

It *should* go without saying by now, but we will say it again. When creating these projects, always follow proper and safe woodworking procedures. This means following the safety guidelines in this book, the safety directions that came with all your tools, and your own common sense.

All right, *now* you can go right ahead and browse through the enticing projects in this section. (All the projects are also pictured in color in the section of photos that follows page 86.) Pick a project, and before too long the following will happen: Someone will visit your house, notice something new, and say, "Where'd that nice bookcase [or table, or cabinet] come from?" You'll then say, as casually as you can manage, "That? Oh, I made it." You'll find that the surprised and admiring response you get will be one of the best rewards that comes from becoming involved in woodworking.

Bathroom Shelf Unit

This shelving unit is a great beginner's project. It's made from easy-to-work pine, and you can complete it with just two power tools—a jigsaw and a drill. (A table saw or a circular saw will make it easier to crosscut all the boards to length and rip a few to the proper width.) The joinery is all simple butt joints. The only challenge in putting the shelves together will be making sure that you've acquired enough clamps in your shop—as you may know already, having too many clamps is a mathematical impossibility.

Materials List

Ref.	Qty.	Item	Dimensions	Material
A	2	Sides	¾ × 6½ × 32	Pine
B	1	Top rail	¾ × 9½ × 18	Pine
C	1	Top shelf	¾ × 5½ × 18	Pine
D	2	Mid and bottom shelves	¾ × 5¼ × 18	Pine
E	1	Bottom rail	¾ × 7 × 18	Pine
F	1	Towel bar	⅝ dia. × 18	Dowel

Other materials: No. 8 × 1½" wood screws

1 Once you have the boards cut to length, mark the profiles for the side pieces. Using the drawing on the facing page as a guide, you could mark the shape directly onto one board and cut it out with your jigsaw (if you are feeling confident). A more precise and safer method is to draw the shape on cardboard first and cut it out to be sure it looks okay, then trace it onto the board.

After you have cut out one side piece, use it as a guide to trace the curve on the other side piece (or use the cardboard pattern, flipped over). This will guarantee that the two side pieces will have the exact same curves.

2 Next draw the wavy shapes for the 18"-long top and bottom rails. To achieve symmetry in your design, take a 9"-wide piece of paper or cardboard and draw a curvy line on it. Cut out the design and trace it onto one half of the 18"-long board. Then flip over the cardboard and trace the design onto the other half of the board. Repeat the process for the bottom border, using a different curve (as shown in the drawing).

When you're riding your jigsaw around the curves, guide the blade smoothly so you don't create bumps and dings that you'll have to sand out later *(see photo)*. Many jigsaws have a dial that allows you to adjust the stroke of the blade to be either gentle or aggressive, depending on the wood you're cutting (and the mood you're in).

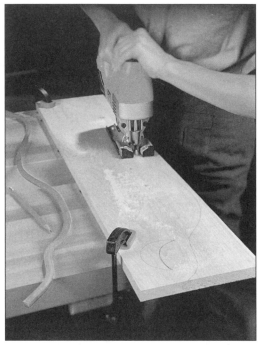

Step 2

3 Next, cut out the crescent moon on the side pieces. (If you wish, you can skip this decorative detail.) Draw the shape where you want it (making sure it won't be in the same spot as the top rail or the top shelf) and drill a generous hole in the middle of the moon. Insert your jigsaw blade into that hole, and proceed to cut out the moon. You should definitely use a scrolling blade for this purpose. A scrolling blade is more delicate than a regular jigsaw blade and can handle tight corners that would make a standard blade buck.

4 Once you've cut out all your pieces, round over the sharp edges with sandpaper. You could instead use a cornering tool to ease all the straight edges. A Veritas cornering tool makes short work of softening the edges of pine *(see photo)*. If the blade chips or binds, you're working against the grain. Try pulling from the opposite end of the board. Grain lines always curve around knots, so you may have to change direction several times, but it's still faster than sanding.

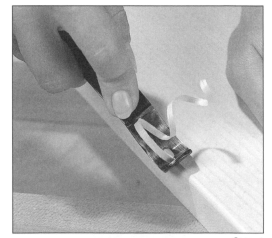

Step 4

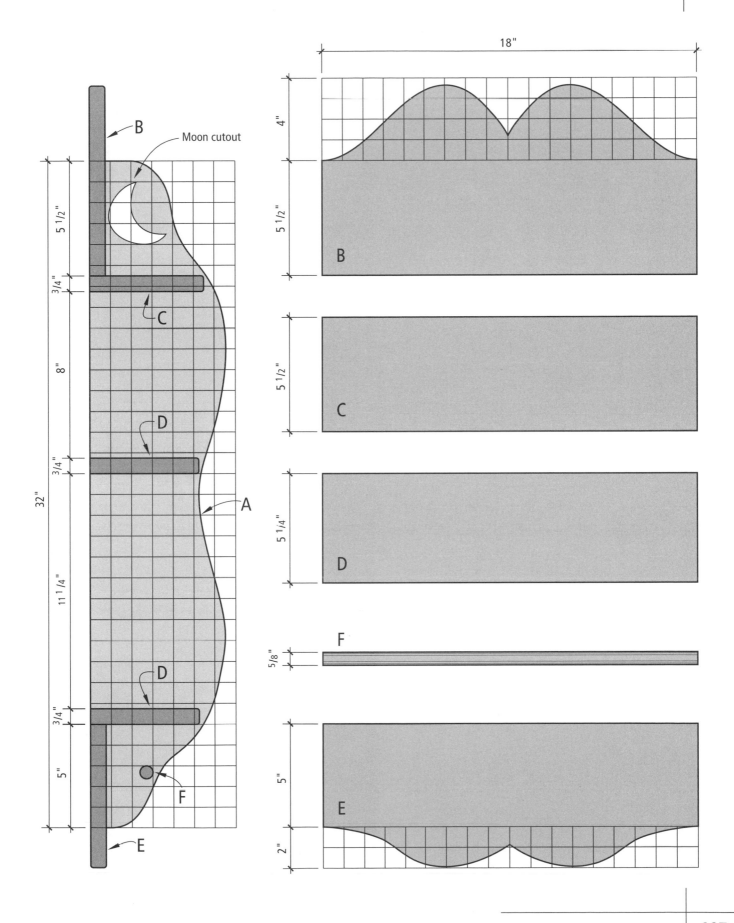

18"

4"

5 1/2"

B

B

5 1/2"

C

C

5 1/4"

D

D

F

5/8"

F

5"

E

E

2"

Moon cutout

B

5 1/2"

3/4"

C

8"

D

3/4"

A

32"

11 1/4"

D

3/4"

5"

F

E

5 It's tempting to screw the whole unit together right now, but there are several advantages to putting the stain, paint, or clear-coat on the individual pieces before you go any further. It is much easier than having to cover all the multiple surfaces of the shelf unit after it's assembled. Besides that, glue squeeze-out will glom on to bare wood during the assembly process, but if your boards already sport a coat of finish, the glue can easily be wiped off.

Step 5

The piece pictured here has been finished with shellac. Shellac has to be applied with a patient, steady hand in smooth, long strokes that don't overlap *(see photo above)*. Because it's alcohol-based, it dries extremely fast, so there's little downtime. If you're mixing your own shellac, use a good solvent procured from a reliable woodworking supply store.

Even if you don't work with shellac as your final finish, you can brush a couple of quick dabs of shellac on any knots. This will seal the knots and prevent sap from oozing up under your chosen finish. In fact, shellac is a great primer coat for pretty much any finish except stain.

6 Now you're ready to "dry fit" the whole unit. Lay it all out with the sides, shelves, and borders in place. Clamp everything together lightly and square the shelves using a speed square. Make those clamps nice and tight. Use a pencil to mark a light line on the side pieces under each of the shelves for reference later when you're doing the final assembly under the duress of knowing that the glue is starting to set up.

7 While all the clamps are in place, predrill pilot holes for your screws so you don't split the shelves when you drive the screws. Then, right on top of the predrilled screw holes, drill larger holes to a depth of ⅜" to make a cavity for the plugs that will hide the screw heads. To avoid drilling the plug holes too deep, it's a good idea to wrap a piece of masking tape around your drill bit ⅜" from the end so you know when to stop *(see photo)*.

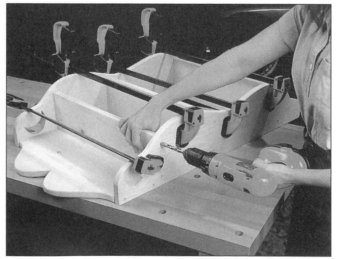

Step 7

8 Now for the "wet fit." Take all the pieces apart and apply a modest bead of glue along the edges of the shelves and borders, plus a dab on each end of the towel bar. Reassemble and clamp everything together, and drive those screws. Use a damp rag to wipe away any oozing glue around the joints.

9 To make plugs to cover the screw heads, you can use a plug-cutter bit. This is a device that fits into your drill just like a regular bit. It cuts tiny cylindrical wood plugs that camouflage the screw heads, so the finished project looks tidy.

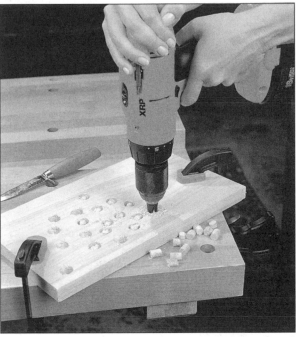

Once your screws are in place, fire up the plug-cutter and cut the plugs in a scrap piece of pine *(see photo)*. Once you've cut about thirty plugs, use a knife or screwdriver to pop each plug out of its little hole.

As an alternative to cutting your own plugs, you can buy precut hardwood plugs at the hardware store. This may be easier, but it has some downsides: the color of the wood won't match, and hardwood isn't as absorbent as pine so the plugs take finish differently and won't blend in.

Step 9

10 Before inserting each plug into its predrilled hole, place a drop of glue on the bottom of each plug and smear it around a bit.

When you have your unit plugged, wait twenty minutes for the glue to set up. Then use a flush-cut saw to cut each plug flush with the surface of the cabinet. If you don't have a flush-cut saw, you can take the plugs down fairly quickly using a sander loaded with 80-grit sandpaper. Once the plugs are cut and sanded, touch up the plugs and sides of your unit with shellac or whatever finish you're using.

11 Now it's time to mount your unit on the wall. One mounting technique is to screw two 2"-long strips of metal plumber's tape (which has prepunched holes) on the back of the shelf unit, then lift the unit onto waiting nails anchored in studs above the toilet. Finding the studs is a matter of importance, because you don't want your unit to come crashing down on you at an unfortunate moment.

Handy Box

This toolbox is perfect for holding all those everyday tools that you use around the house—a tack hammer, screwdrivers, tape ruler, extra screws and nails, etc.

The techniques used to make this box are simple and fun. The miter joints that are used here will require a table saw to cut all the 45° angles (a circular saw could be used, but would be much more difficult). Because of the large area of the beveled joint, a glue bond is sufficient to make a strong connection here.

A clear finish was used in the pictured project in order to keep the wood visible. You can finish this project—and the others in the book—as you like, with paint, stain, or whatever suits your needs.

Materials List					
Ref	Qty.	Item	Dimensions	Material	Notes
A	2	Top and bottom	¾ × 6 × 16	Plywood	All edges have a 45° bevel
B	2	Front and back	¾ × 7½ × 16	Plywood	All edges have a 45° bevel
C	2	Ends	¾ × 6 × 7½	Plywood	All edges have a 45° bevel

Other materials: 1 continuous hinge (³⁄₁₆" × 1½" × 16"); 1 draw bolt; 1 screen door handle

1 Cut all the parts to size as shown in the Materials List. Then tilt the table saw blade to 45° and attach a sacrificial fence to the saw's fence. Adjust this setup until you can cut a 45° bevel on the edge of the box parts. This setup allows you to cut bevels on parts that have already been cut to size, which is easier than trying to cut all the parts to size and bevel them at the same time.

A safety note: Be careful when cutting and handling plywood after you've cut bevels on the edges. Plywood edges are fragile and can easily be chipped or nicked. Also, you could receive some nasty cuts from these edges because they are sharp.

Step 1

2 Double-check the 45° bevel to be sure that a perfect 90° corner is formed. This is critical for all the parts to join together squarely at glue-up time.

3 Lay out the bottom A, the front and back B, and the two ends C face up as shown. Use clear packing tape to tape the joints, creating a hinge. Be sure that the sharp edges of the bevels come together as cleanly as possible when you apply the tape.

4 Turn the whole assembly face down, then apply glue to all the edges that will be coming together when it's folded up.

Step 2

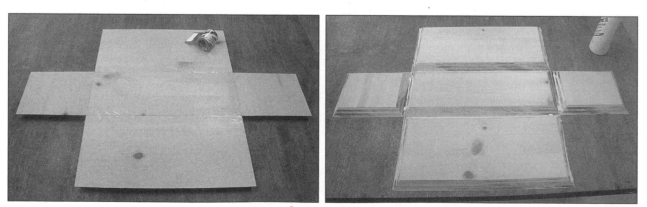

Step 3

Step 4

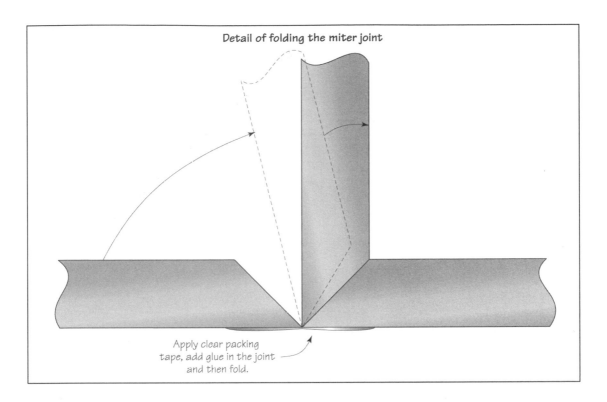

Detail of folding the miter joint

Apply clear packing
tape, add glue in the joint
and then fold.

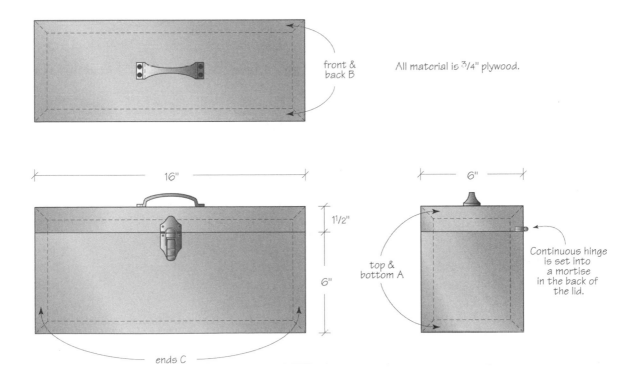

front &
back B

All material is ³/4" plywood.

16"

1¹/2"

6"

ends C

6"

top &
bottom A

Continuous hinge
is set into
a mortise
in the back of
the lid.

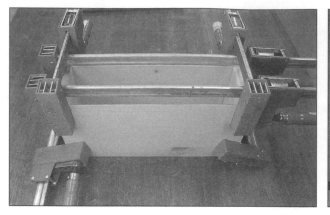

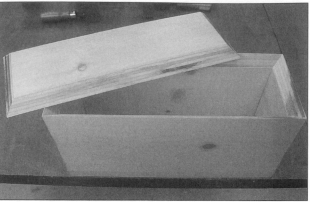

5 You will need to use a few clamps to hold the side joints tightly. Don't use too much pressure, as that will distort the joint and cause it to open up at the sharp edges.

6 When the glue has dried, remove the clamps, apply glue to the remaining beveled edges, and attach the top.

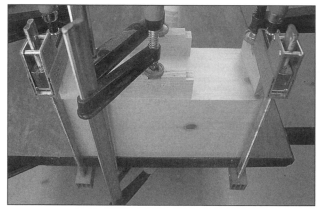

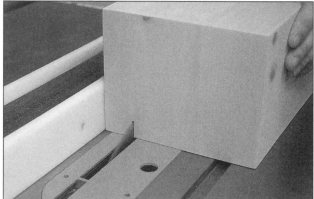

7 Use blocks under the clamps to even out the pressure. At this point you might be wondering if you could tape the top to the assembly when all the parts are lying flat in Step 3. Yes, you could!

8 When the glue has dried, gently scrape or sand away any glue squeeze-out. Cut the lid off the box, using the table saw.

9 The lid will fit perfectly on the box using this technique.

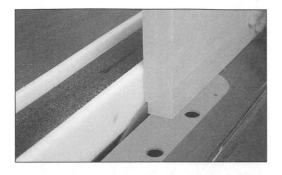

10 Measure the thickness of the continuous hinge. Set the table saw fence to this measurement and make a through-cut.

11 Reset the fence to the width of the hinge leaf (this does not include the barrel of the hinge) and make the cleanup cut. This cut squares out the corner of the rabbet cut.

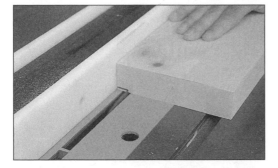

12 Install the hinge. If your cut is accurate, the hinge is lined up easily by holding the edge of the hinge leaf against the shoulder of the rabbet. By allowing the barrel of the hinge to extend beyond the edge of the box, the lid can be opened 180°.

13 This is a quick but neat and tidy way to install a hinge. You can now install the draw bolt and screen door handle.

14 When you've finished the box, the corners will have a nice, clean look. This is a strong joint. The gluing surface is large, and no splines or biscuits are needed.

Desktop Organizer

No matter how big your desk is, it's never quite big enough for all the things you want to put on top of it, is it? This project will help you create a place to store a few things, therefore keeping your desktop a little neater. As shown here, the organizer is made of curly maple, which makes the piece visually attractive. The basics of biscuit joinery are all that you will need to build this project. The shelf unit is assembled with biscuits, as is the drawer. Before final assembly, all the parts should be sanded and finished.

Materials List				
Ref.	Qty.	Item	Dimensions	Material
A	1	Top	¾ × 10 × 30	Maple
B	2	Sides	¾ × 9¾ × 17¼	Maple
C	3	Shelves	¾ × 8¾ × 26½	Maple
D	1	Back	¾ × 14¼ × 27½	Maple
E	1	Drawer front	¾ × 3 × 26½	Maple
F	2	Drawer sides	½ × 3 × 7¾	Maple
G	1	Drawer back	½ × 2¼ × 25½	Maple
H	1	Drawer bottom	¼ × 8 × 26	Luan plywood
J	2	Pulls	¾ dia. × 1	Oak dowel

Other materials: Biscuits

1 Lay the two side panels side by side and cut all the biscuit slots at the same time. This will ensure that the shelves will be level and the project will be square when assembled.

Step 1

2 Lay the shelves face up on a flat surface when cutting the slots. The bottoms of the shelves will then be located correctly with the slots that have been cut in the sides.

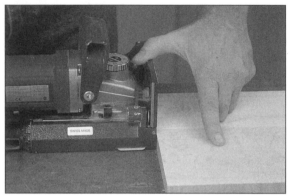

Step 2

3 Stand the top on its back edge and cut the slots that line up with the back panel. Next, cut the slots in the top where the sides will attach to the top, using the layout square. To locate the slots, dry assemble the base of the project and measure the distance between the slots in the tops of the sides. Transfer this measurement to the top panel, making sure the top will be centered on the base.

4 After the slots have been cut and you've determined that the parts fit together properly, finish sand all the parts. Cover the slots with masking tape, and finish the parts.

Step 4

Step 3

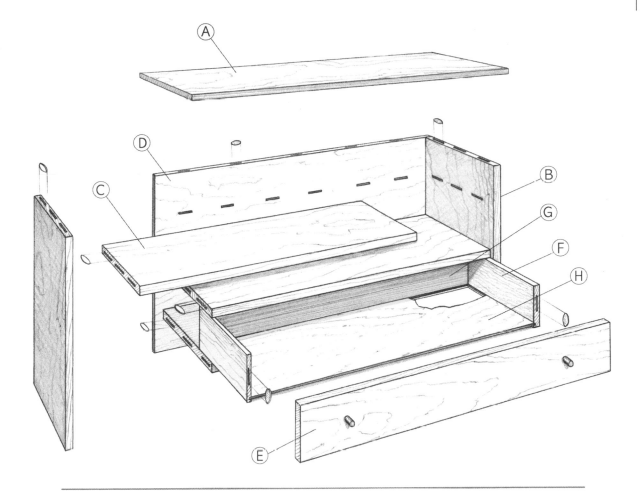

5 To keep the finish out of the slots, cover them with tape.

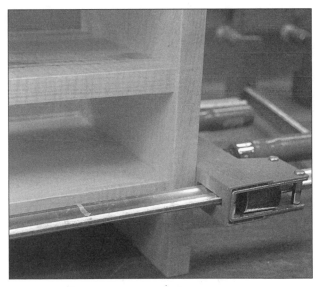

6 Let the finish cure overnight, then assemble the project. (You will find it easier to clean up any glue that oozes out of the joints, because the unit was finished prior to assembly.)

7 The back is screwed to the base. Note that the screw is inserted at a slight angle so it won't cause a bulge on the inside of the project.

Step 7

8 Glue the top in place.

Step 8

Step 9

9 To make the drawer using biscuit joinery: Lay out all the drawer parts in order, and label using an orienting triangle. Remember that the drawer sides will capture the front and back when the drawer is assembled.

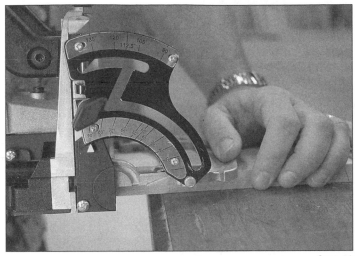

Step 10

10 Mark the location of the biscuits and cut the slots in the front and back parts. When making drawers, use the largest biscuits you can.

11 Cut the slots into the sides.

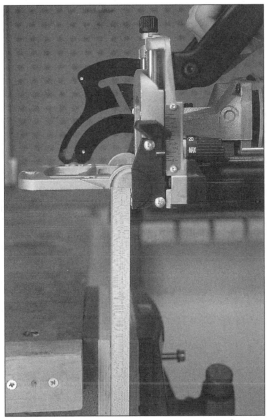

Step 11

12 Assemble the drawer. You can either use a captured bottom (as shown here), where the bottom is held in dadoes on all four sides. Or you cut off the bottom of the back part; the bottom can then be slid into the assembled drawer after the drawer and bottom have been finished.

Step 12

Step 13

13 To finish up, drill holes for the white oak ¾" dowel pulls. Any wooden dowel would make a nice, simple pull for the drawer.

Hanging Cupboard

This small cupboard is a reproduction of an original Shaker design created in Mount Lebanon, New York, in the early 1800s; it was created by scaling a photograph of the piece to create a measured drawing. It has two adjustable shelves and would look good in any room in the house.

This is a good project for learning and using fundamental biscuit joinery techniques. The original Shaker cupboard in the photograph was made of pine and had a warm brown color. As shown here, the cabinet is made of sugar pine and finished with orange shellac. Then it was topcoated with precatalyzed lacquer to provide a water-resistant finish.

Materials List

Ref.	Qty.	Item	Dimensions	Material
A	1	Back	½ × 11½ × 16	Sugar pine
B	2	Sides	½ × 4½ × 13	Sugar pine
C	2	Fronts	½ × 2 × 13	Sugar pine
D	2	Inner top and bottom	½ × 4½ × 10½	Sugar pine
E	2	Outer top and bottom	½ × 5¾ × 13	Sugar pine
F	1	Door	½ × 7½ × 13	Sugar pine
G	1	Latch	¼ × ¾ × 1½	Sugar pine

Other materials: 1 pair brass hinges (⅜" × 1"); wooden knob; biscuits

1 After cutting all parts to size, cut the slots for the ends and sides. Dry-assemble the parts before gluing to be sure all the joints fit properly.

2 When clamping the sides and top and bottom together, be sure the assembly is square.

3 Here the slots are being cut into the two face panels. By laying the two parts on edge together, it's easier to hold the biscuit joiner square to the face of the panels.

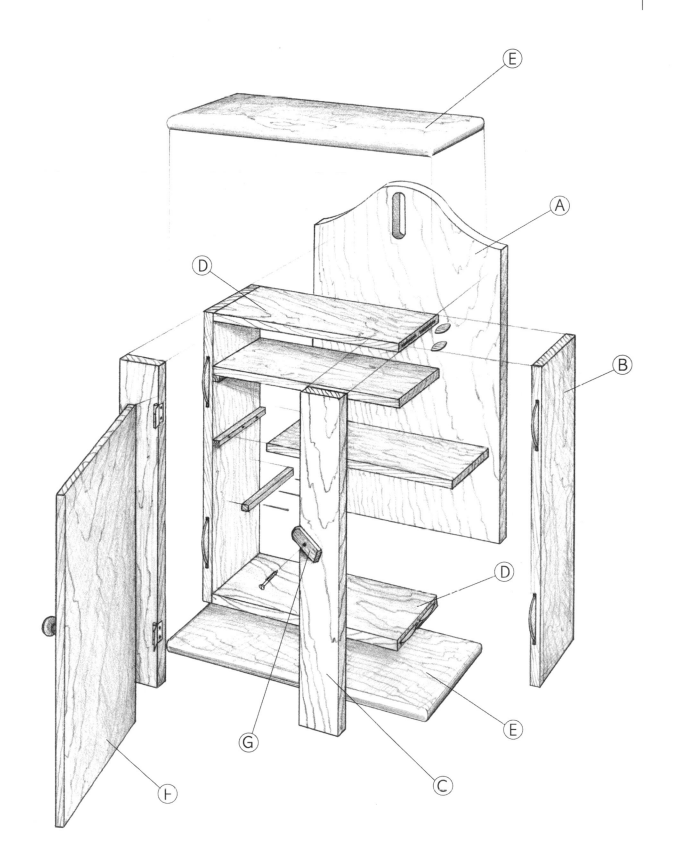

E

A

D

B

D

E

G

C

F

4 Two biscuits will help align the panels with the cabinet sides.

5 Be sure to use scrap pieces of wood between the clamps and the pieces being glued. This sugar pine is easily dented by the clamps.

6 Attach the shelf brackets with small nails or brads. Use a spot of glue in the middle of the bracket if desired.

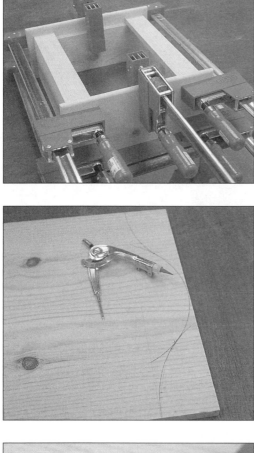

7 Glue the top and bottom outer panels to the cabinet.

8 To lay out the arch on the top of the back panel, first draw the center radius. Then, blend a second radius into the center radius to form a smooth transition. Cut and sand this curve smooth.

9 Drill oversize screw holes (¼") in the back panel and attach it to the cabinet with screws. The oversize holes will let the back panel expand and contract with changes in humidity without cracking or splitting.

10 The hinges are mortised into the door and screwed directly to the edges of the front panels. You can then attach the knob to the front door and the door to the hinges.

Small Tool Chest

This small tool chest will sit nicely on a bench but can easily be moved. It comes with a removable box, handy for holding those small screws and bits that always seem to get lost.

The frames are joined with tongue-and-groove joints; these are similar to the mortise-and-tenon joints also commonly used in frame-and-panel construction. Besides being a good introduction to this type of joinery, the project as pictured here is a testament to the value of recycling. All of the frame elements were made from one board of pallet wood, and the panels were cut from scrap pieces of ¼" luan plywood.

Materials List				
Ref.	**Qty.**	**Item**	**Dimensions**	**Material**
A	2	Top and bottom	¾ × 8¼ × 20½	Hardwood
B	8	Stiles	¾ × 1¾ × 9¼	Hardwood
C	4	Side rails	¾ × 1¾ × 5½	Hardwood
D	4	Front and back rails	¾ × 1¾ × 17½	Hardwood
E	2	Side panels	¼ × 6¾ × 5½	Plywood
F	2	Front and back panels	¼ × 6¾ × 17½	Plywood
Removable Box:				
G	2	Box sides	½ × 1½ × 8	Plywood
H	2	Box ends	½ × 1½ × 5⅜	Hardwood
J	2	Box runners	½ × 3 × 18½	Hardwood
K	1	Box bottom	¼ × 6⅜ × 8	Plywood

Other materials: 1 continuous hinge (³⁄₁₆" × 1½" × 20")

1 Cut out the parts as listed in the Materials List. Use parts B, C, D, E, and F to build the front, back, and side panels as shown in "How to Assemble a Frame and Panel" on pages 132 and 133. Then cut a 45° bevel on all the ends of the panels.

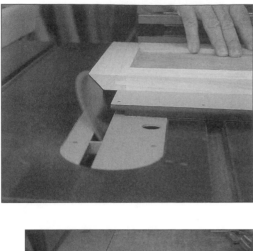

2 Lay the panels face up and end to end. Tape the joints with clear packing tape. If you like, press the tape flat with a steel or wooden roller.

3 Flip the whole assembly face down, apply glue to the bevel joints, and fold it together.

4 Use clamps to hold the untaped corner joint until the glue dries. Be sure the assembly is square. Cut the rabbet in the top of the back panel, attach the continuous hinge, and install the top A. Attach the bottom A with screws and install the runners J for the removable box.

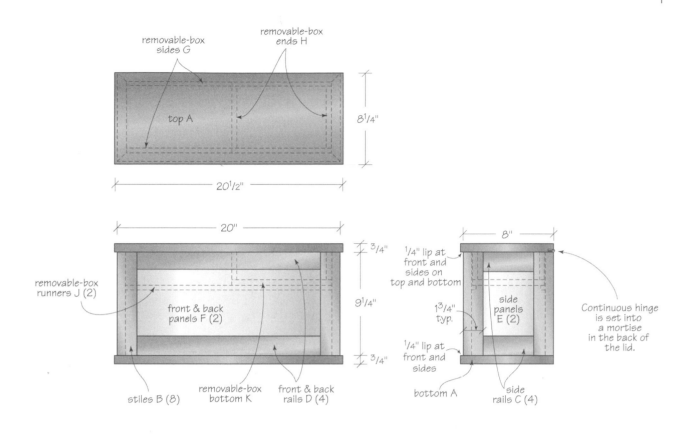

removable-box sides G

removable-box ends H

top A

8¹/₄"

20¹/₂"

20"

3/4"

9¹/₄"

3/4"

removable-box runners J (2)

front & back panels F (2)

stiles B (8)

removable-box bottom K

front & back rails D (4)

¹/₄" lip at front and sides on top and bottom

1³/₄" typ.

side panels E (2)

¹/₄" lip at front and sides

8"

Continuous hinge is set into a mortise in the back of the lid.

bottom A

side rails C (4)

5 Cut out the removable-box parts G, H, and K and glue the front, back, and sides together using butt joints.

6 After the glue dries, glue the bottom K to the removable box assembly. Sand the whole project and finish. Glued butt joints are strong enough for this box because the bottom helps hold the whole assembly together.

How to Assemble a Frame and Panel

Frame and panels are commonly used in cabinetry, particularly for making doors. Here's what you need to know to put them together. You may notice, by the way, that the solid-wood panel pictured here has a bevel along all its edges. This allows the panel edges to fit into the narrower grooves of the frame. In the tool chest project, the panels are already all ¼"-thick plywood, so using this technique is not necessary.

1 Lay out all of the parts as they will be in the final assembly. Without this step, it is easy to make mistakes—for example, turning a panel upside down.

2 Add glue to the end of the first rail (horizontal piece) and set it in place on the stile (vertical piece).

3 Put the panel in place.

4 Add glue to the other rail and put it in place.

5 Put glue on the ends of both rails and put the other stile in place.

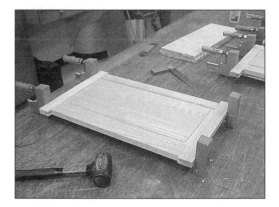

6 Lay the assembly flat on the clamps; apply slight pressure with the clamps. Check the assembly for squareness, adjust if necessary, and add a little more pressure until the joints are tight. Don't apply any more pressure than is necessary, as it could distort the joint.

Sliding-Door Cabinet

As pictured here, this wall-mounted cabinet has holes drilled through the top two shelves to store router bits. Without the holes, this project could provide storage in all sorts of places—a workshop, a garage, or even a small bathroom.

This piece looks great as it is made here in ash, with poplar shelves, but this is one project where the choice of woods is really up to you. The wooden handles also look great—and work well for sliding the doors in both directions—but they do require several steps to make. You may want to substitute simpler handles, possibly from the hardware store.

Materials List

Ref.	Qty.	Item	Dimensions	Material
A	2	Sides	¾ × 5 × 16	Ash
B	2	Top and bottom	¾ × 5 × 24⅜	Ash
C	1	Back	¾ × 14½ × 24⅜	Plywood
D	2	Sliding doors	⅜ × 11½ × 15	Hardwood
E	2	Handles	⅞ × 1 × 15	Ash
F	2	Shelves	½ × 3⅜ × 24⁵⁄₁₆	Poplar

Other materials: Biscuits

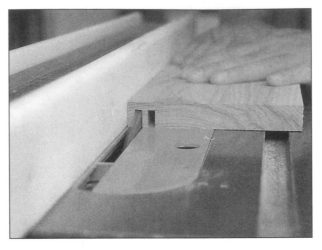

1 Cut the parts to size. Then cut the grooves for the sliding doors in the top and bottom (B). The grooves in the bottom are ¼" deep.

2 The grooves in the top are ½" deep. After the cabinet is assembled, the doors can be lifted up into these grooves and then lowered into the grooves in the bottom panel.

3 Glue the top and bottom panels to the back panel (C). Note the use of spacers to help hold the assembly square.

4 Cut biscuit slots in the sides (A).

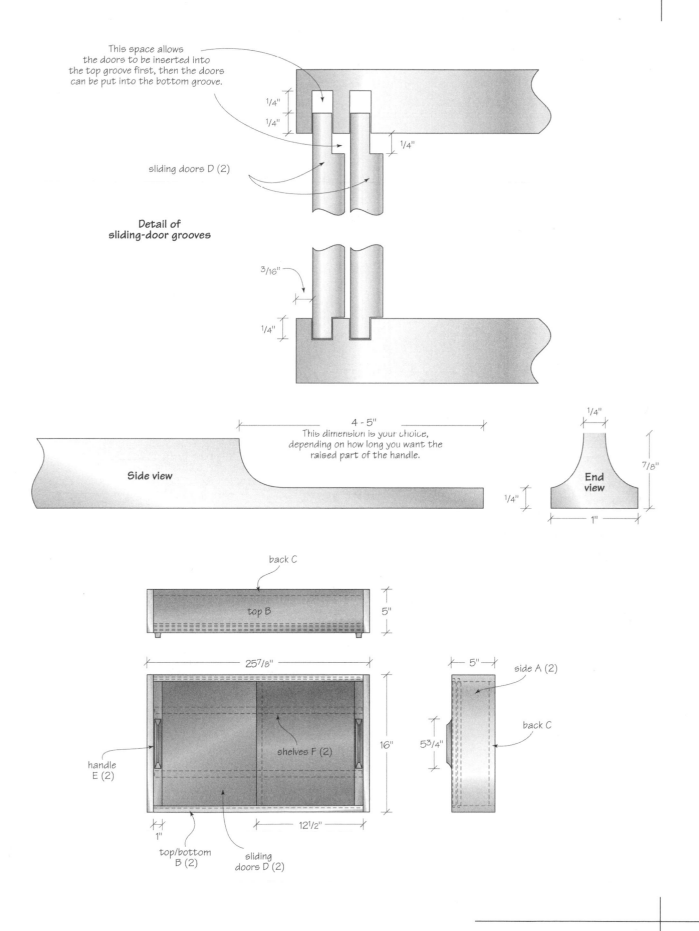

This space allows the doors to be inserted into the top groove first, then the doors can be put into the bottom groove.

1/4"

1/4"

1/4"

sliding doors D (2)

Detail of sliding-door grooves

3/16"

1/4"

4 - 5"
This dimension is your choice, depending on how long you want the raised part of the handle.

Side view

1/4"

1/4"

7/8"

End view

1"

back C

top B

5"

25⁷/₈"

16"

shelves F (2)

handle E (2)

1"

top/bottom B (2)

sliding doors D (2)

5"

side A (2)

back C

5³/₄"

5 Cut matching biscuit slots in the ends of the top and bottom panels.

6 Glue the sides to the cabinet.

7 The handles (E) for the doors are made using a ½" cove cutter on the router table. Make the cove cuts on a larger piece of stock.

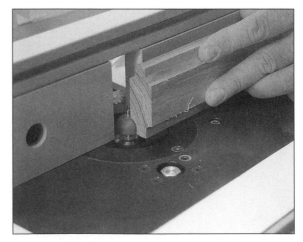

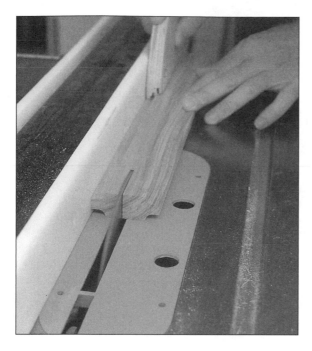

8 Cut the handles apart on the table saw.

9 Set up a straight-cutting bit on the router table. Mount a stop block to the fence and use it to start the offset cut on the handle. Flip the handle around to make the other cut.

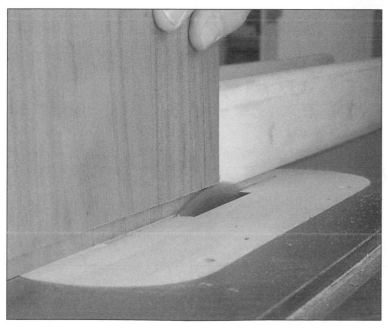

10 Glue the handles to the door panels (D). Then cut the tenons on the doors to fit into the grooves in the top and bottom of the cabinet. Be sure the door slides freely. Cut the shelves (F) to size. Lay out all of your router bits (if that's what you're going to store in this cabinet) and drill holes in the shelves to hold the bits in place.

Tall Bookcase with Drawers

This project is based on actual plans for a Craftsman bookcase. The original has tenons on its shelves that go through the sides and are held in place with wedges inserted into the ends of the tenons. The shape, dimensions, and hardware have been kept from the original plans. Using biscuits, this bookcase was much easier to make—and it looks great. Instead of the traditional Arts and Crafts dark staining, the white oak in this piece has a clear finish. The wood will eventually develop a beautiful golden-brown patina.

Materials List				
Ref.	Qty.	Item	Dimensions	Material
A	2	Sides	¾ × 14 × 66	Oak
B	3	Top and two bottoms	¾ × 13 × 34½	Oak
C	1	Center divider	¾ × 6 × 12⅞	Oak
D	1	Kick plate	¾ × 2 × 34½	Oak
E	1	Top shelf rail	¾ × 1½ × 34½	Oak
F	3	Shelves	¾ × 10 × 34½	Oak
G	1	Back	¾ × 34½ × 65	Oak
Back Parts:				
H	2	Stiles	¾ × 4 × 65	Oak
J	1	Crest rail	¾ × 7 × 26½	Oak
K	1	Bottom rail	¾ × 12 × 26½	Oak
L	2	Panels	¼ × 12 × 47	Oak
M	1	Center stile	¾ × 4 × 46	Oak

Other materials: 12 shelf pins; biscuits

1 After cutting all the parts, lay out the sides next to each other so it is clear to you which will be left and right. Use a set of trammel points to draw the arc at the tops of the sides. Bending a thin strip of wood to the desired arc will also work.

2 Here is a method for drawing the curves at the bottom and the tops of the sides that will produce arcs with consistent, pleasing shapes. (You may choose to instead draw these curves in a simpler way, as long as the general dimensions are the same.)

First, at the bottom of the sides, mark how wide the bottom of the feet will be and draw these vertical lines using a square held against the bottom of the side panels. Mark how far up the cutout will go, and mark that square to the sides. Then draw another line parallel to this line ¼" below it. Use a cup or small can and draw an arc connecting the vertical lines with the lower horizontal line. Bend a thin strip of wood (or a small-diameter wooden dowel rod) connecting the small arcs with the top horizontal line at its center. Assuming that you only have two hands, you'll need to have someone help you draw this arc.

3 Mark the arc on the crest rail as shown. Mark on both ends of the rail how far down you would like the arc to go. Put a mark at the center top of the rail, and connect the dots as shown.

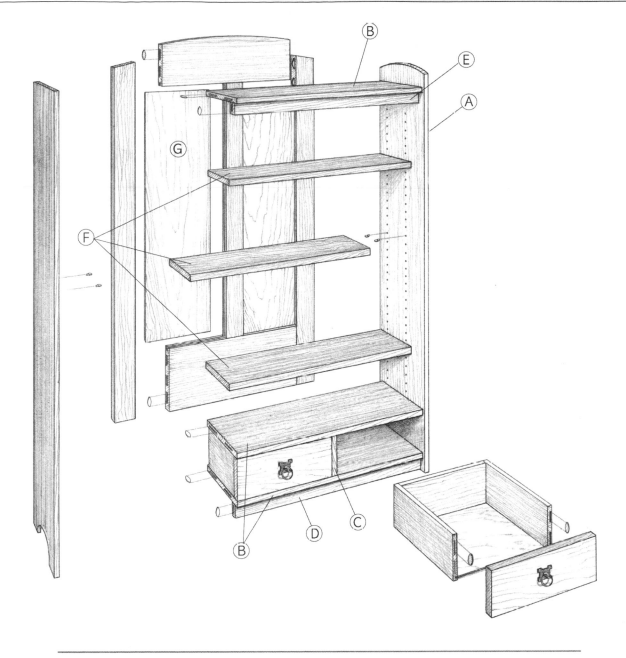

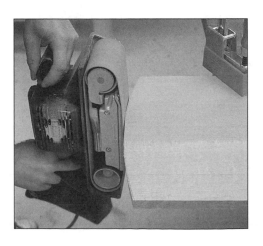

4 Rough-cut the top arcs, and smooth them with a belt sander or by hand with a sanding block.

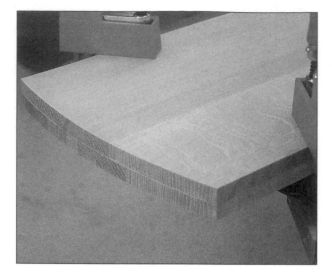

5 You could also make a template and rout the arcs. This ensures that both sides will be the same, and it also makes the final edge cleanup easier.

6 Cut the bottom arcs and sand them smooth, or make a template and rout the cutouts.

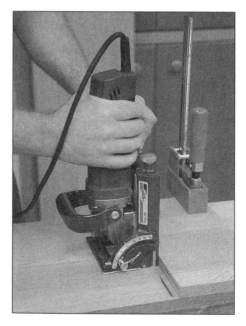

7 Cut the first set of slots in the side panels.

8 Turn the biscuit cutter around and cut the second set of slots.

9 Draw a line parallel to and ⅜" from the front edge of both the top and bottom panels. Glue the kick rail and top rail in place, using the lines as your reference.

10 A router set up with a ¼" wing cutter is a good way to make the stopped grooves in the back panel rails and stiles.

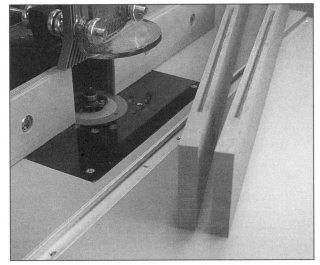

11 Before assembling the back panel, finish sand the ¼" panels. After the back panel is assembled, sand the entire project and finish it before final assembly. You'll be gluing this bookcase together in one operation, which includes twenty-two biscuits and a dozen clamps, so gather all the necessary materials before final assembly and have them within reach.

12 Working quickly, apply the glue to all the parts and clamp them all together. You might need to call back that assistant who helped you draw the curves earlier to help hold the back panel in place and lend a hand attaching the clamps.

Occasional Table

There is space and need in almost everyone's home for an occasional table. The attractive design—in particular, the tapered legs—of this project make it a good choice to fill that need. To further sweeten the deal, its design has a dual purpose. Every family room has those certain items you need only occasionally (is that where the name came from?), but there's never a good place to keep them. You know, the remote control you rarely use, the pocket dictionary, or the coasters for when company's around. Well, lift off the top of this table and you've uncovered a storage space for those occasionally needed items.

Materials List				
Ref.	Qty.	Item	Dimensions	Material
A	4	Legs	1¾ × 1¾ × 26	Cherry
B	2	Aprons	⅞ × 3½ × 17½	Cherry
C	2	Aprons	⅞ × 3½ × 21½	Cherry
D	1	Top	¾ × 20½ × 24½	Ash
E	1	Bottom	¼ × 20½ × 24½	Plywood

Other materials: Biscuits; No. 6 × ¾" flat screws

1 Construction begins by cutting out the parts according to the Materials List. Start with the tapered legs. There are many methods for doing this, but the simplest is just laying out the taper on each leg, cutting it out with a band saw, and planing the taper with a bench plane.

First determine which sides of the legs will face out, choosing the best figure for those faces. The tapers are only on the two inside faces of each leg. To keep the legs correctly oriented, place the legs as they will be on the finished table, then hold them together and mark a diamond across the intersection of all four legs at the top (photo).

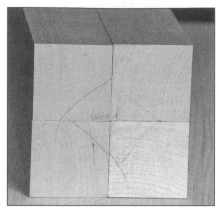

Step 1

2 Next, mark each leg on the inside face (where the aprons will butt against them) at 4⁵⁄₁₆" and 4¹³⁄₁₆" down from the top. The 4⁵⁄₁₆" measurement is the location of the bottom edge of the apron, which leaves ¹⁄₁₆" of the leg protruding above the top, adding a nice detail. The 4¹³⁄₁₆" measurement is the starting location of the leg taper.

3 Now move to the bottoms of the legs and using a combination square, mark a 1" square on each leg, measuring from the outside corner. This indicates where the inside tapers will end on each leg. Connect the marks from the top to the bottom of the legs; then cut the tapers on a band saw, cutting as close to the line as you can (photo). To smooth out the band saw cut, use a bench plane and a bit of muscle to remove the rough-sawn edge.

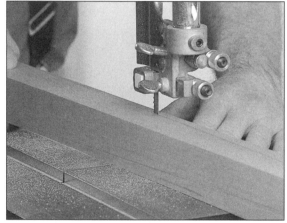

Step 3

4 The term *occasional table* implies that this table won't be expected to carry a lot of weight. In that spirit, the joinery doesn't have to be extraordinarily strong. Two No. 20 biscuits in each joint provide plenty of strength for the base. The ¼" plywood bottom screwed in place will add to the base's strength.

Set up the joinery by marking each leg 2⁹⁄₁₆" from the top (the center point for the aprons). Adjust the biscuit joiner to space two biscuits evenly in the thickness of each apron and to position the aprons flush to the legs.

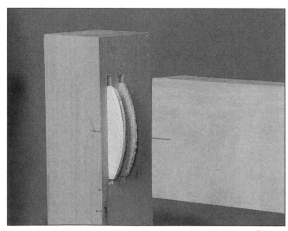

Step 4

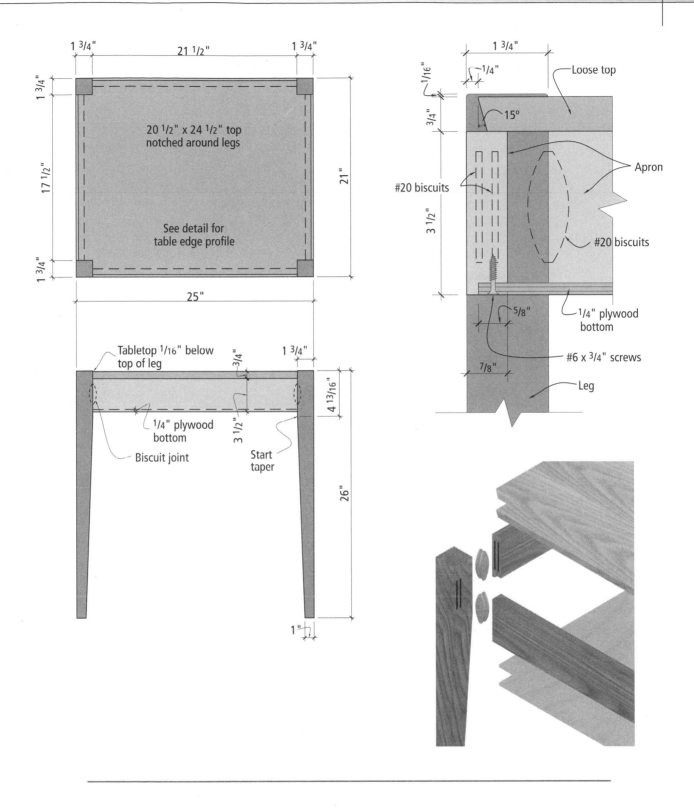

1 3/4" **21 1/2"** **1 3/4"**

1 3/4"

17 1/2"

20 1/2" x 24 1/2" top
notched around legs

21"

See detail for
table edge profile

1 3/4"

25"

Loose top

1 3/4"

1/16"

1/4"

15°

3/4"

#20 biscuits

Apron

3 1/2"

#20 biscuits

1/4" plywood
bottom

5/8"

#6 x 3/4" screws

7/8"

Leg

Tabletop 1/16" below
top of leg

3/4"

1 3/4"

1/4" plywood
bottom

3 1/2"

Biscuit joint

Start
taper

4 13/16"

26"

1"

5 After cutting the biscuit joints, set up a router to run a ¼" × ⅝"-wide rabbet in the bottom edge of the aprons for the bottom. With the rabbets cut, start assembling the base by gluing the short aprons between the legs. Dealing with fewer clamps on any procedure makes the glue-up easier.

6 Check for squareness on each glued-up end by measuring from the top corner of one leg to the bottom corner of the other, making sure the measurements are equal. After about an hour, glue up the rest of the base, again checking the base for squareness on the sides and across the width and length of the apron. For the loose top to fit accurately, you have to be on the money.

Take the time to wipe off any glue that you see before it dries. The inside of the table gets finished, so you have to keep squeeze-out to a minimum.

7 Cut out a ¼" bottom to fit the dimensions between the rabbets in the aprons. To make the bottom fit in place correctly, notch the corners around the legs using the band saw. Don't install the bottom until after finishing.

8 The last construction step is the top. The top's width was achieved by gluing up four thinner boards. After the top is glued up and dry, cut it to the same size as the outside dimensions of the table base, which is a bit bigger than the finished size of the top.

9 Using the table base as a reference helps you cut accurate notches in the top. Mark the notch locations by laying the top upside down on a clean surface, then turn the base upside down and lay it on the top, flushing the corners. Mark the leg locations for the notches.

To notch the top using a table saw, clamp a ¾" spacer board to the rip fence about 3" back from the leading edge of the blade. Set the blade's height and the distance from the fence (including the blade) to the size of your notch and add ¹⁄₁₆" to the cut to allow room for wood movement.

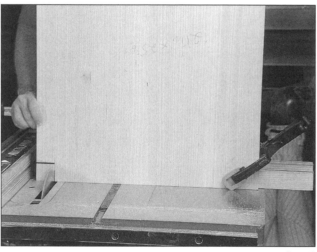

Step 9

10 Run the top on edge against the saw's miter gauge. It's a good idea to add a sacrificial board to the miter gauge as well to add some extra height for support, and to back the top behind the notch to reduce tear-out. Push the top up to the spacer block on the fence, then push it past the blade, holding the top tightly against the miter gauge. The spacer block allows you to properly align the piece for the cut, but keeps the notch (once cut free) from binding between the blade and fence, which could cause a dangerous kickback. Check the fit of your top. To allow you to lift the top, it needs to be a little loose. Next, cut the bevels on the top's edges by setting your table saw's blade to 15°. Set the rip fence so the cut is almost flush to the top edge of the top, leaving about a ¹⁄₃₂" flat on the edge. This cut will remove about ¼" off the underside of the edge. Repeat this cut on the other edges, then finish sand the top.

11 After sanding the base, you're ready to apply a finish. The project as shown here was finished with a mix of boiled linseed oil and stain. The top gets no stain. Apply three coats of clear finish to the base, top, and bottom. When the final coat is dry, screw in the bottom with some No. 6 × ¾" flat-head screws.

Glossary of Woodworking Words and Phrases

A

adhesive A substance that holds materials together by surface attachment. It's a general term that includes a variety of substances, including cement, mucilage, paste, and glue.

adhesive joint The location at which two elements are held together with a layer of adhesive.

American Lumber Standard This establishes standard sizes and requirements for the development and coordination of lumber grades of various species, and the implementation of those standards through an accreditation and certification program, set by the American Softwood Lumber Standard.

assembly joint Joints between variously shaped parts or subassemblies in wood furniture.

B

back-priming Applying a coat of primer to the back of a cabinet door panel to prevent warping.

batten A thin, narrow strip of (usually) plywood that conceals a joint between adjoining pieces of lumber or plywood.

bevel A cut that is not 90° to a board's face (that is, one at some other angle, often 45°), or the face left by such a cut.

bird's-eye Small, localized areas in wood with the fibers indented and contorted to form circular or elliptical figures on the surface; this pattern is often found in sugar maple.

biscuit A thin, flat oval of compressed beech that is inserted between two pieces of wood into mating grooves made by a biscuit or plate-joining machine.

board Lumber that is less than 2" nominal thickness and greater than 2" nominal in width. Boards less than 6" nominal width are sometimes called strips.

board foot A unit of measurement of lumber represented by a board that is 12" long, 12" wide, and 1" thick, or the cubic equivalent of those measurements.

bond Basically, to glue together; veneers are "bonded" to create a sheet of plywood.

bow Distortion of a structural wood panel so that it is not flat lengthwise.

bridle joint A joint that combines features of both lap joints and mortise-and-tenon joints. It has a U-shaped mortise in the end of the board.

burl The hard, woody outgrowth on a tree, often used for highly figured, decorative veneers.

butt joint The joint formed when two pieces of wood are fastened together without overlapping.

C

carpenter's glue White and yellow adhesives formulated for use with wood.

casing The trim framing a window, door, or other opening.

caulk Waterproof sealant used to fill joints or seams; available as putty, a rope, or a compound squeezed from a cartridge.

chalk line Line made by snapping a chalk-coated string against a plane.

chamfer The flat surface created by slicing off the square edge or corner of a piece of wood or a panel.

check A separation between growth rings at the end of a board. Checks are common and lessen appearance, but do not weaken wood unless deep.

chipboard A paperboard with a variety of uses.

cleavage The separation in a joint caused by a wedge in an adhesively bonded joint.

close-grained wood Wood with narrow, inconspicuous annual rings; often designates wood with small pores.

coarse-grained wood Wood with wide, conspicuous annual rings; often designates wood with large pores.

composite panel A veneer-faced panel with a reconstituted wood core.

compound miter A cut where the blade path is not perpendicular to the wood's end or edge and the blade tilt is not 90° to the face.

coping Sawing a negative profile in one piece to fit the positive profile of another, usually in molding.

corbel A projection from the face of a wall or column supporting a weight.

core The inner ply (or plies) whose grain runs perpendicular to that of the outer plies; or, a layer of reconstituted wood.

counterbore A straight-sided drilled hole that recesses a screw head below the wood surface so a wood plug can cover it, or the bit that makes this hole.

countersink A cone-shaped drilled hole whose slope angle matches the underside of a flat screw head and sinks it flush with the wood surface, or the tool that makes this hole.

crook An end-to-end warp along the board edge.

crossband The veneer layers with grain direction perpendicular to that of the face plies in plywood.

crosscutting Sawing wood across the grain. Because the wood in structural wood panels like plywood is either cross-laminated or randomly oriented, any cut made in a structural wood panel is a crosscut. Always use a crosscut saw when hand or power sawing structural wood panels.

cross-grained wood   Wood in which the fibers deviate from a line parallel to the sides of the piece.

cup   Crosswise distortion from the flat plane of a structural wood panel.

curly-grained wood   Wood in which the fibers are distorted so they have a curled appearance, as in bird's-eye maple.

D

dado joint   Joint formed by the intersection of two boards, one of which is notched with a rectangular groove.

decay   The decomposition of wood caused by fungi.

decorative panel   An interior or exterior plywood panel grade with rough-sawn, brushed, grooved, or striated faces.

delamination   The separation of layers in laminated wood or plywood due to adhesive failure.

density   Weight per unit volume. Density of wood is influenced by the rate of growth and the percentage of late wood.

dimension   Lumber with a thickness from 2" nominal up to, but not including, 5" nominal and a width of greater than 2" nominal.

dovetail joint   A traditional joint characterized by interlocking fingers and pockets shaped like its name. It has exceptional resistance to tension.

dowel   A small cylinder of wood that is used to reinforce a wood joint.

dressed size   Dimensions of lumber after being surfaced with a planing machine, usually ½" to ¾" less than the nominal (or rough) size.

dressing   The process of turning rough lumber into a smooth board with flat, parallel faces and straight, parallel edges and whose edges are square to the face.

dry rot   This term refers to any dry, crumbly wood rot, which usually causes the wood to become powdery.

E

edge-grained lumber   Wood that has been sawed so the wide surfaces extend at about right angles to the annual growth rings.

edge joint   Joint made by bonding two pieces of wood together edge to edge, usually with glue; may be made by gluing two squared edges or by machined joints such as tongue-and-groove joints.

edge lap   A notch into the edge of a board halfway across its width that forms half of an edge lap joint.

end grain   The end of a piece of wood exposed when the wood fibers are cut across the grain, at right angles to the direction of the fibers.

end joint   Joint made by bonding two pieces of wood together end to end, often by a finger or scarf joint.

exterior plywood   Plywood bonded with an adhesive that is resistant to the effects of weather.

153

extruded particleboard Particleboard made by shoving particles into a heated die, forming a rigid mass.

F

face The highest-grade side of any veneer-faced panel that has outer plies of different veneer grades.

fascia Wood or plywood trim used along the eave or the gable end of a structure.

fiberboard Sheet materials of a variety of densities that are manufactured from refined or partially refined wood fibers.

fiddleback-grained wood Often used for violin backs, this wood figure is produced by a type of fine, wavy grain found in maple and other wood species.

figure The pattern produced in a wood surface by annual growth rings, rays, knots, or deviations from regular grain.

finger joint End joint made up of several meshing wedges or fingers of wood bonded together with adhesive.

fingerlap This specific joint of the lap family has straight, interwoven fingers; also called a box joint.

finish Varnish, stain, paint, or any mixture that protects a surface; also refers to fine woodwork needed to complete a project, especially a building's interior.

flakeboard Particle panel product composed of wood flakes.

flat-sawn The most common cut of lumber, where the growth rings run predominantly across the end of the board; or its characteristic grain pattern.

flush Level with an adjoining surface.

foam core Center of a plywood "sandwich" panel. Liquid plastic foamed into all spaces between the plywood panels insulates and supports the component skins.

G

glue Now synonymous with "adhesive," glue originally referred to a hard gelatin made of hides, tendons, cartilage, bones, and other animal parts, and the adhesive created from this substance by heating it with water.

glue line The adhesive joint formed between veneers in a plywood panel or between face veneers and core in a composite panel.

grain The natural growth pattern in wood. The grain runs lengthwise in the tree and is strongest in that direction.

grain pattern The visual appearance of wood grain. Types of grain pattern include flat, straight, curly, quilted, rowed, mottled, crotch, cathedral, bee's-wing, or bird's-eye.

group number Plywood is manufactured from more than seventy species of softwood that are classified according to strength and stiffness into Groups 1 through 5. Group 1 woods are the strongest. The group number of a particular panel is determined by the weakest (highest numbered) species used in face and back (except for some thin panels where strength parallel to face grain is unimportant).

growth ring A tree's annual cross-sectional growth layer, including springwood and summerwood.

gum A sticky accumulation of resin that bleeds through finishes.

H

hardboard Panel manufactured from mostly wood, consolidated under heat.

hardwood Wood of the deciduous or broad-leaved trees—such as oak, maple, ash, or walnut. Hardwood is only a general term, not a reference to actual wood hardness.

heartwood The nonactive core of a tree distinguishable from the growing sapwood by its usually darker color and greater resistance to rot and decay.

I

interior plywood Plywood manufactured for indoor use.

interlocked grain Grain in which the tree's wood fibers may slope in a right-handed direction for several years, then reverse to a left-handed direction for several years, and so on.

J

jig A shop-made or aftermarket device that assists in positioning and steadying the wood or tools.

jointing The process of making a board face straight and flat or an edge straight, whether by hand or machine.

K

kerf A slot made by a saw, or the width of a saw cut.

key An inserted joint-locking device, usually made of wood.

kiln-dried Wood dried in ovens, or kilns, by controlled heat and humidity to specified limits of moisture content.

knife marks The imprints or markings of the machine knives on the surface of dressed lumber.

knockdown joint A joint that is assembled without glue and can be disassembled and reassembled if necessary.

knot Natural growth characteristic of wood caused by a branch base imbedded in the tree trunk.

knothole The void that is produced when a knot drops out of veneer.

L

laminated veneer lumber (LVL) Structural wood elements constructed of veneers laminated together with their fibers oriented in a parallel direction.

lap To position adjacent objects so that one surface extends over the other.

lap joint Joint made by placing one member partly over another and bonding the overlapped portions.

length joint A joint that makes one longer wood unit out of two shorter ones by joining them end to end.

lumber The wood product of a sawmill and/or planing mill, with all four sides sawn and/or planed.

lumber core Plywood manufactured with a core composed of lumber strips. The face and back, or outer, plies are veneer.

M

machine burn A burn on the face of a board sometimes caused by blunt planer knives.

machine wave   Waves on the face of wood caused by incorrect planer speeds. Boards with waves must be thinned again.

medium-density fiberboard (MDF)   Panel product manufactured mostly from wood fibers combined with synthetic resin or other binding agents.

milling   The process of removing material to leave a desired positive or negative profile in the wood.

miter   A generic term mainly meaning an angled cut across the face grain; or specifically a 45-degree cut across the face, end grain, or along the grain. See also *bevel*.

molding   A wood strip that has a curved or projected surface, used for decorative purposes.

mortise   The commonly rectangular or round pocket into which a mating tenon is inserted. Mortises can be blind (stop inside the wood thickness), through, or open on one end.

N

nominal size   As applied to lumber, the size by which it is known and sold in the market, which often differs from the actual size.

O

oriented strand board (OSB)   Particle panel product composed of strandlike flakes purposefully aligned in directions to make a panel stronger and stiffer.

overlay   A thin layer of paper, plastic, film, metal foil, or other material bonded to one or both faces of a piece of lumber or panel product to provide a protective or decorative face or a base for painting.

P

pallet   A low wood or metal platform on which material can be stacked to facilitate mechanical handling, moving, or storage.

paperboard   A substance that is thicker than paper, heavier, and more rigid.

particleboard   A panel made of wood particles and glue.

peeler log   A specially selected softwood log used to produce veneer.

pilot hole   A small, drilled hole used as a guide and pressure relief for screw insertion, or to locate additional drilling work like countersinking and counterboring.

pitch pockets   Pitch-filled spaces between grain layers that may bleed after the board is milled; occasionally bleeds through finishes.

pith flecks   Irregular, discolored streaks of tissue in wood, due to insect attacks on the growing tree.

ply   A single veneer in a panel.

plywood   Panel made by laminating layers of wood.

pressure-preservative treated   Wood treated with preservatives or fire retardants; treating solutions are pressure-injected into wood cells.

primer   An undercoat applied to bare wood as a sealer and base for paint.

Q

quartersawn   A stable lumber cut where the growth rings on the board's end run more vertically across the end than horizontally and the grain on the face looks straight; also called straight-grained or rift sawn.

R

rabbet joint A joint formed by cutting a groove into the surface or along the edge of a board, plank, or panel to receive another piece.

rail A horizontal part of a frame.

raised grain A roughened condition of the surface of dressed lumber in which the hard summerwood is raised above the softer springwood, but not torn loose from it.

resawn lumber The product of sawing any thickness of lumber to develop thinner lumber.

ripping Sawing wood in the direction of the grain.

rough lumber Lumber that has not been dressed or surfaced, but has been sawed, edged, and trimmed.

S

sapwood The living wood of a tree, often an off-white, pale color, located near the outside of a log, outside the heartwood. Usually, sapwood is more susceptible to decay than heartwood.

scarf joint An angled or beveled end joint splicing pieces together.

scribe To make layout lines or index marks using a knife or awl.

seasoning Evaporation or extraction of moisture from green or partially dried wood.

shake A separation between growth rings that results in a slat coming loose from the face of the board.

shiplap Jointing in which ends or edges are notch-milled to overlap and form a rabbet joint.

shoulder The perpendicular face of a step cut, like a rabbet, that bears against a mating joint part to stabilize the joint.

slope of grain The deviation of the line of fibers from a straight line parallel to the sides of the piece.

softwood Wood of the coniferous or needle-leaved trees—such as pine, fir, spruce, or hemlock. Softwood is only a term, not a general reference to actual wood hardness.

spline A flat, thin strip of wood that fits into mating grooves between two parts to reinforce the joint between them.

springwood The portion of the annual ring formed during the early part of the yearly growth period of a tree. Lighter in color, less dense, and not as strong as summerwood.

stain A pigment or dye used to color wood through saturation; or, a discoloration in wood from fungus or chemicals.

steam-bending The process of forming curved wood members by steaming or boiling the wood and bending it to a form.

stile A vertical part of a door frame.

summerwood The portion of the annual ring formed during the latter part of the yearly growth of a tree. Darker in color, more dense, and stronger than springwood.

surfaced lumber Lumber that has been dressed by running it through a planer.

T

tenon The male part of a mortise-and-tenon joint, commonly rectangular or round; it projects as the wood is cut away around it so it can be inserted into a mortise.

texture   Determined by relative size and distribution of the wood elements. Described as coarse (large elements), fine (small elements), or even (uniform size of elements).

tongue-and-groove joint   A system of jointing in which the rib or tongue of one member fits exactly into the groove of another.

trim   To crosscut a piece to any given length.

twist   A lopsided or uneven warp. Wood is weakened, but twisted boards are fit for non-load-bearing use.

V

veneer   A thin sheet of wood laminated with others under heat and pressure to form plywood, or used for faces of composite panels. Also referred to as ply.

W

waferboard   Panels manufactured from reconstituted wood wafers, as opposed to strands, bonded with resins under heat and pressure like oriented strand board (OSB).

warp   The distortion in lumber that causes a departure from its original plane, usually developed during drying. Warp includes cupping, bowing, crooking, and twisting.

width joint   A joint that makes a unit of the parts by joining them edge to edge to increase the overall width of wood.

wood-based composite panel   A generic term referring to material manufactured from wood veneer, strands, flakes, particles, or fibers mixed with a synthetic resin or other bonding agent.

Woodworking Resources

The following listing of books and Web sites are here for you to further your education of the woodworking craft. Of course, you can always ask for help and advice of those who work at your local woodworking shop and home improvement store.

Web Sites

www.popularwoodworking.com
The go-to site for any question, tool, material, or project idea you need.

www.woodworking.com
For amateurs and experts alike, this is an excellent resource for woodworking information.

www.rockler.com
This manufacturer's site is a good place to learn something new that you didn't know about woodworking.

http://woodworking.about.com
This site offers a plethora of information on woodcraft, especially for beginners.

www.woodmagazine.com
This magazine's site is an excellent woodworking resource.

Books

Popular Woodworking Complete Book of Tips, Tricks, and Techniques
Use this text to help you get the most out of woodworking projects. Turn to it for useful advice and expertise from people who know everything there is about wood.

The Insider's Guide to Buying Tools: The Bottom Line for the Best Tool Values
It's not as easy as you think. Purchasing tools that are appropriate for the projects you want to create is of utmost importance.

The Best of Wood Boxes
From basic to advanced, these projects are good for collecting anything from old memorabilia to bed linens.

The Essential Pine Book
Many woodworkers, especially those just starting out in the hobby, love pine and want to make everything from it. If that sounds like you, stop right here and pick up this all-about-pine book.

Suppliers

Adams & Kennedy—the Wood Source
6178 Mitch Owen Road
P.O. Box 700
Manotick, Ontario
Canada K4M 1A6
613-822-6800
www.wood-source.com
Wood supplier

Adjustable Clamp Company
417 North Ashland Avenue
Chicago, Illinois 60622
312-666-0640
www.adjustableclamp.com
Clamps and woodworking tools

Delta Machinery
4825 Highway 45 North
P.O. Box 2468
Jackson, Tennessee 38302-2468
800-223-7278 (U.S.)
800-463-3582 (Canada)
www.deltawoodworking.com
Woodworking tools

Exaktor Tools, Ltd.
136 Watline, #182
Mississaugua, Ontario
Canada L4C 2E2
800-387-9789
www.exaktortools.com
Sliding tables and other accessories for
the table saw

General and General International
8360 du Champ-d'Eau
Montreal, Quebec
Canada H1P 1Y3
514-326-1161
www.general.ca
Woodworking machinery

House of Tools, Ltd.
100 Mayfield Common Northwest
Edmonton, Alberta
Canada T5P 4B3
800-661-3987
www.houseoftools.com
Woodworking tools and hardware

JessEm Tool Company
124 Big Bay Point Road
Barrie, Ontario
Canada L4N 9B4
866-272-7492
www.jessem.com
Rout-R-Slide and Rout-R-Lift

Langevin & Forest
9995 Pie IX Boulevard
Montreal, Quebec
Canada H1Z 3X1
800-889-2060
www.langevinforest.com
Tools, hardware, and lumber

Lee Valley Tools, Ltd.
P.O. Box 1780
Ogdensburg, New York 13669-6780
800-871-8158 (U.S.)
800-267-8767 (Canada)
www.leevalley.com
Woodworking tools and hardware

LRH Enterprises, Inc.
9250 Independence Avenue
Chatsworth, California 91311
800-423-2544 (U.S.)
818-782-0226 (outside U.S.)
www.lrhent.com
Router bits and the Magic Molder

Porter-Cable
4825 Highway 45 North
P.O. Box 2468
Jackson, Tennessee 38302-2468
800-321-9443
www.porter-cable.com
Woodworking tools

Richelieu Hardware
7900, West Henri-Bourassa
Ville St-Laurent, Quebec
Canada H4S 1V4
800-619-5446 (U.S.)
800-361-6000 (Canada)
www.richelieu.com
Hardware supplies

Rockler Woodworking and Hardware
4365 Willow Drive
Medina, Minnesota 55340
800-279-4441
www.rockler.com
Woodworking tools and hardware

Trend Machinery & Cutting Tools, Ltd.
Odhams Trading Estate
St. Albans Road
Watford
Hertfordshire, U.K.
WD24 7TR
01923 224657
www.trendmachinery.co.uk
Woodworking tools and hardware

Vaughan & Bushnell Mfg. Co.
11414 Maple Avenue
Hebron, Illinois 60034
815-648-2446
www.vaughanmfg.com
Hammers and other tools

Wolfcraft North America
333 Swift Road
Addison, Illinois 60601-1448
630-773-4777
www.wolfcraft.com
Woodworking hardware

Woodcraft
P.O. Box 1686
Parkersburg, West Virginia 26102-1686
800-535-4482
www.woodcraft.com
Woodworking hardware

Woodworker's Hardware
P.O. Box 180
Sauk Rapids, Minnesota 56379-0180
800-383-0130
www.wwhardware.com
Woodworking hardware

**Plywood and Particleboard Material
Information and Suppliers**
www.panolam.com
www.uniboard.com

index

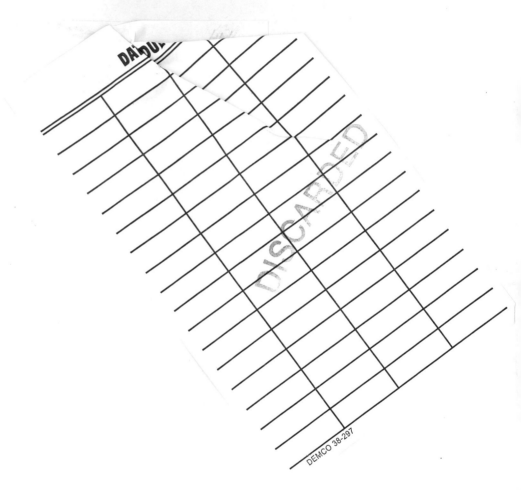